Market Timing and Moving Averages

Market Timing and Moving Averages

An Empirical Analysis of Performance in Asset Allocation

Paskalis Glabadanidis

palgrave
macmillan

MARKET TIMING AND MOVING AVERAGES

First published in 2015 by
PALGRAVE MACMILLAN®
in the United States—a division of St. Martins Press LLC,
175 Fifth Avenue, New York, NY 10010.

Where this book is distributed in the UK, Europe and the rest of the world, this is by Palgrave Macmillan, a division of Macmillan Publishers Limited, registered in England, company number 785998, of Houndmills, Basingstoke, Hampshire RG21 6XS.

Palgrave Macmillan is the global academic imprint of the above companies and has companies and representatives throughout the world.

ISBN: 978–1–137–36468–5

Library of Congress Cataloging-in-Publication Data

Glabadanidis, Paskalis.
Market timing and moving averages : an empirical analysis of performance in asset allocation / Paskalis Glabadanidis.
pages cm
Includes bibliographical references and index.
ISBN 978–1–137–36468–5 (hardback : alk. paper)
1. Technical analysis (Investment analysis) 2. Stock price indexes. 3. Portfolio management. I. Title.

HG4529.G58 2015
332.63'2042–dc23 2015001256

A catalogue record of the book is available from the British Library.

Design by Newgen Knowledge Works (P) Ltd., Chennai, India.

First edition: July 2015

10 9 8 7 6 5 4 3 2 1

To Z. because timing is everything.

Contents

Figures

Tables

Acknowledgments

I would like to thank Syed Zamin Ali, Tze Chuan 'Chewie' Ang, B. Ross Barmish, Jean Canil, Don Chance, Sudipto Dasgupta, Daisy Doan, Victor Fang, Mebane Faber, Berowne Hlavaty, Daniel Orlovsky, James Primbs, Bruce Rosser, Vincent Xiang, Takeshi Yamada, Alfred Yawson, Xinwei Zheng, Edward Zychowicz as well as seminar participants at Deakin University and the University of Adelaide and participants in the 2012 Australasian Finance and Banking conference in Sydney, the 2014 J.P. Morgan quantitative conference in Sydney and the 2013 Midwest Finance Association meetings in Chicago. Any remaining errors are my own responsibility.

Chapter 1

Fundamental Versus Technical Analysis

Technical analysis involves the use of past and current market price, trading volume, and, potentially, other publicly available information to try and predict future market prices. It is highly popular in practice with plentiful financial trading advice that is based largely, if not exclusively, on technical indicators. In a perhaps belated testament to this fact consider the following quote from the *New York Times* issue dated March 11, 1988: "*Starting today the New York Times will publish a comprehensive three-column market chart every Saturday... History has shown that when the S&P index rises decisively above its (moving) average the market is likely to continue on an upward trend. When it is below the average that is a bearish signal.*" More formally, Brock, Lakonishok and LeBaron (1992) find evidence that some technical indicators do have a significant predictive ability. Blume, Easley and O'Hara (1994) present a theoretical framework using trading volume and price data leading to technical analysis being a part of a trader's learning process. A more thorough study of a large set of technical indicators by Lo, Mamaysky and Wang (2000) also found some predictive ability especially when moving averages (MAs) are concerned. Zhu and Zhou (2009) provide a solid theoretical reason why technical indicators could be a potentially useful state variable in an environment where investors need to learn over time the fundamental value of the risky asset they invest in. More recently, Neely, Rapach, Tu and Zhou (2010, 2011) find that technical analysis has as much forecasting power over the equity risk premium as the information provided by economic fundamentals. The practitioners literature also includes Faber (2007) and Kilgallen (2012) who thoroughly document the risk-adjusted returns to the MA strategy using various portfolios, commodities, and currencies. In addition, Huang and Zhou (2013) use the MA indicator to predict the US stock market while Goh et al. (2013) apply the same idea to government bond risk yields and premia. Motivated in part by the predictive power of the MA indicator, Han and Zhou (2013) and Jiang (2013) construct a trend factor with considerable cross-sectional explanatory power and substantial historical performance.

The main findings of this study are as follows. First, I present evidence that the returns to a simple MA switching strategy dominate in a mean-variance sense the returns to a buy-and-hold strategy of the underlying portfolio. Second, I demonstrate that the switching strategy involves infrequent trading with relatively long periods when the MA strategy is invested in the underlying assets and the break-even transaction costs (BETC) are on the order of 0.05% to 0.25% per transaction. Thirdly, even though there is overwhelming evidence of imperfect market-timing ability of the MA switching strategy for portfolios, that is not the case for individual stocks. Furthermore, cross-sectional differences remain between the portfolio and individual stock abnormal returns. These differences persist when controlling for the four-factor Carhart (1997) model for portfolios formed on past price returns and are mostly driven by differences in the volatility of portfolio and stock returns. Fourthly, conditional models explain to a certain degree the MA abnormal returns but do not completely eliminate them. Fifth, I show that the lagged indicator regarding the switch into or out of the risky asset has substantial predictive ability over subsequent portfolio returns over and above the predictability contained in standard instrumental variables, like the stock market's dividend yield, the short-term risk-free rate, and a recession dummy variable.

This study is similar in spirit to that of Han, Yang and Zhou (2013). However, several important differences stand out. First, I use daily value-weighted returns of decile portfolios constructed by various characteristics like size, book-to-market, momentum, past returns, and industry classification.[1] Value-weighted portfolios at a daily frequency should have a smaller amount of trading going on inside the portfolio compared to the daily equal-weighted portfolios investigated by Han, Yang and Zhou (2013). Secondly, the cross-sectional results in this study are just an artifact of the decile portfolios and not the main focus of this paper while the study by Han, Yang and Zhou (2013) is mostly concerned with the inability of standard empirical tests to account for the MA strategy average returns differences across portfolios. I argue that this is largely due to using the wrong benchmark pricing model. Using a dynamic market-timing tests and conditional asset pricing models with macroeconomic state variables leads mostly negative or statistically insignificant risk-adjusted returns for the MA strategy. In light of this, my take on the performance of the MA strategy is that it is not an anomaly but instead a dynamic trading strategy that exposes investors to potential upside returns derived from risky assets via its market-timing ability. This performance is more pronounced the more volatile the returns of the underlying risky assets are. A final caveat I need to make is that the performance of the strategy is investigated using historical returns rather than actually trading in financial markets. It is likely that in reality there may be adverse price impact of liquidating and initiating large positions, especially for less liquid assets with lower trading volumes. This possibility is in the spirit of limits to arbitrage as another potential explanation for the performance of the MA strategy. The nature of this empirical study is such that this potential explanation cannot be eliminated.

The highlights of this study are the superior performance of the MA portfolios relative to buying and holding the underlying portfolios, the infrequency of trading and positive BETC, the fact that the switching strategy returns resemble an imperfect at-the-money protective put, and that cross-sectional differences are not a new anomaly as maintained in Han, Yang and Zhou (2013) but are due to volatility differences in the underlying portfolios and stocks. An asset with 10% higher standard deviation of returns will experience on average between 1.6% and 3.5% mean return improvement between the buy-and-hold and the MA strategy. The returns of the MA strategy relative to the buy-and-hold strategy exhibit a lot of convexity and, hence, will be hard to explain using standard linear asset pricing models. The anomalous risk-adjusted performance relative to standard models appears to be largely due to omitting market-timing factors in a simple piece-wise linear framework that captures the MA strategy's convexity. Furthermore, the MA strategy appears to be antifragile in the sense of Taleb (2012), meaning that for securities with more volatile returns there is a greater improvement of the moving average returns relative to buy-and-hold returns.

Chapter 2

Investment Performance

In this chapter I use daily value-weighted[1] returns of sets of ten portfolios sorted by market value, book-to-market, momentum, short-term and long-term price reversal, and industry classification. The data is readily available from Ken French Data Library. The sample period starts on January 4, 1960, and ends on December 31, 2013.

The following exposition of the moving average (MA) strategy follows closely the presentation in Han, Yang and Zhou (2013). Let R_{jt} be the return on portfolio j at the end of day t and let P_{jt} be the respective price level of that portfolio. Define the moving average of portfolio (MAP) j $A_{jt,L}$ at time t with length L periods as follows:

$$A_{jt,L} = \frac{P_{jt-L+1} + P_{jt-L+2} + \cdots + P_{jt-1} + P_{jt}}{L}, \tag{2.1}$$

Throughout most of the chapter, I use an MA of length $L = 20$ days. My reasons for selecting an MA length of $L = 20$ days are threefold.[2] First, I want to keep the MA length using a daily frequency similar to the MA length used by Han, Yang and Zhout (2013). Secondly, I am interested in using a reasonably long length of the MA in order to avoid excessive trading which would have eroded the performance of the strategy. Thirdly, I only report my findings for $L = 20$ because I would like to avoid any objections concerning potential data mining. In the first run of the experiment I used $L = 20$ and I will continue reporting all empirical findings for $L = 20$. Later on, in the section dealing with robustness checks I also present results for all sets of portfolios with lags of 5 days, 10 days, 60 days, 120 days, and 250 days. According to Brock, Lakonishok and LeBaron (1992), the MA in its various implementations is the most popular strategy followed by investors who use technical analysis. The way I implement the MA strategy in this paper is to compare the closing price P_{jt} at the end of every day to the running daily MA $A_{jt,L}$. If the price is above the MA this triggers a signal to invest (or stay invested if already invested at $t-1$) in the portfolio on the next day $t+1$. If the price is below the MA this triggers a signal to leave the risky portfolio (or stay invested in the risk-free asset if not invested at $t-1$) in the following day $t+1$. As a proxy for the risk-free rate, I use the daily return of the 30-day US Treasury Bill.

More formally, the returns of the MA switching strategy can be expressed as follows:

$$\tilde{R}_{jt,L} = \begin{cases} R_{jt}, & \text{if } P_{jt-1} > A_{jt-1,L} \\ r_{ft}, & \text{otherwise,} \end{cases} \tag{2.2}$$

in the absence of any transaction costs imposed on the switches. The alternative specification for the case of positive one-way transaction cost of τ leads to the following four cases in the post-transaction cost returns:

$$\tilde{R}_{jt,L} = \begin{cases} R_{jt}, & \text{if } P_{jt-1} > A_{jt-1,L} \text{ and } P_{jt-2} > A_{jt-2,L}, \\ R_{jt} - \tau, & \text{if } P_{jt-1} > A_{jt-1,L} \text{ and } P_{jt-2} < A_{jt-2,L}, \\ r_{ft}, & \text{if } P_{jt-1} < A_{jt-1,L} \text{ and } P_{jt-2} < A_{jt-2,L}, \\ r_{ft} - \tau, & \text{if } P_{jt-1} < A_{jt-1,L} \text{ and } P_{jt-2} > A_{jt-2,L}. \end{cases} \tag{2.3}$$

depending on whether the investor switches or not. Note that this imposes a cost on selling and buying the risky portfolio but no cost is imposed on buying and selling the Treasury bill. This is consistent with prior studies like Balduzzi and Lynch (1999), Lynch and Balduzzi (2000), and Han (2006), among others. Regarding the appropriate size of the transaction cost, Balduzi and Lynch (1999) propose using a value between 1 and 50 basis points. Lynch and Balduzi (2000) use a mid-point value of 25 basis points. As a baseline case, I use a zero transaction cost of $\tau = 0$. This leads to a straightforward comparison of the BETC with the actual transaction cost that an investor would experience when trading.

I construct excess returns as zero-cost portfolios that are long the MA switching strategy and short the underlying portfolio to determine the relative performance of the MA strategy against the buy-and-hold strategy. Denote the resulting excess return for portfolio j at the end of day t as follows:

$$\text{MAP}_{jt,L} = \tilde{R}_{jt,L} - R_{jt}, \quad j = 1, \ldots, N, \tag{2.4}$$

The presence of significant abnormal returns can be interpreted as evidence in favor of superiority of the MA switching strategy over the buy-and-hold strategy of the underlying portfolio. Naturally, the MA switching strategy is a dynamic trading strategy so it is perhaps unfair to compare its returns to the buy-and-hold returns of being long the underlying portfolio.

2.1 Profitability of MA Portfolios

In this section, I present summary statistics for the underlying portfolio performance, the performance of the MA switching strategy, and the excess MAP returns for seven sets of ten portfolios sorted by market value, book-to-market ratios, momentum, short-term reversal, long-term reversal, and industry classification. Next, I present single-factor capital asset pricing model (CAPM) of Sharpe (1964), three-factor Fama-French (1992), and four-factor Carhart (1997) regression results for the MAP returns of each set of portfolios. Finally, I discuss the result in light of the potential reasons for the profitability of the MA switching strategy.

2.2 Performance

Table 2.1 reports the first three moments and the Sharpe ratios of the underlying portfolios, the MA switching strategy applied to each portfolio, and the excess return (MAP) of the MA switching strategy over the buying and holding (BH) of the underlying portfolio. The results are most intriguing. First, the average annualized returns of the MA strategy are substantially higher than the average annualized returns of the underlying portfolios. Second, this average return difference comes with a lower return standard deviation and, hence, the MA switching strategy appears to dominate the underlying BH portfolio strategy in a mean-variance sense.[3] Third, for the vast majority of portfolios, the underlying BH has a negative return skewness while the MA strategy in most cases exhibits positive skewness. This feature will make the MA switching strategy very attractive to investors who have a preference for skewness. Fourth, the risk-return trade-off is improved substantially resulting in much higher Sharpe ratios of the MA returns when compared to the Sharpe ratios of the BH returns. Fifth, these results hold for almost all portfolios across all sorting variables.

Table 2.1 Summary statistics

This table reports summary statistics for the respective buy-and-hold (BH) portfolio returns, the MA switching strategy portfolio returns, and the excess return of MA over BH (MAP) using sets of ten portfolios sorted by various characteristics. The sample period covers January 4, 1960 until December 31, 2013 with value-weighted portfolio returns. μ is the annualized average return, σ is the annualized standard deviation of returns, s is the annualized skewness, k is the annualized kurtosis, and SR is the annualized Sharpe ratio. The length of the MA window is 20 days.

Portfolio	BH portfolios					MA portfolios					MAP portfolios				
	μ	σ	s	k	SR	μ	σ	s	k	SR	μ	σ	s	k	SR
						Panel A: Size sorted portfolios									
Low	11.88	13.58	−0.90	14.81	0.52	25.78	8.34	−0.28	12.70	2.52	13.89	10.62	1.52	34.01	1.31
2	11.86	16.72	−0.46	13.85	0.42	21.96	10.53	−0.61	22.32	1.63	10.10	12.94	0.49	28.70	0.78
3	12.96	16.60	−0.49	12.02	0.49	21.56	10.46	−0.15	13.35	1.60	8.60	12.85	0.78	27.31	0.67
4	12.18	16.33	−0.48	12.48	0.45	20.11	10.31	−0.47	16.08	1.49	7.92	12.63	0.62	27.55	0.63
5	12.68	16.23	−0.47	11.96	0.49	19.07	10.31	−0.58	15.95	1.39	6.38	12.51	0.54	26.39	0.51
6	12.26	15.33	−0.53	13.43	0.49	18.53	9.73	−0.51	14.63	1.41	6.27	11.82	0.71	31.06	0.53
7	12.29	15.47	−0.55	15.54	0.49	18.25	9.81	−0.46	15.62	1.37	5.96	11.95	0.79	36.44	0.50
8	12.01	15.70	−0.51	16.62	0.46	16.51	9.99	−0.41	14.97	1.18	4.50	12.10	0.73	39.98	0.37
9	11.57	15.46	−0.59	20.90	0.44	13.70	9.74	−0.18	12.07	0.92	2.14	12.00	1.03	52.10	0.18
High	10.23	16.00	−0.50	21.64	0.34	9.01	10.14	−0.19	13.68	0.42	−1.22	12.38	0.88	54.22	−0.10

Panel B: Book-to-market sorted portfolios															
Low	9.49	17.53	−0.18	13.47	0.27	12.42	10.93	−0.04	10.10	0.70	2.93	13.70	0.26	31.98	0.21
2	11.03	16.07	−0.39	16.37	0.39	12.93	10.24	−0.17	12.80	0.80	1.90	12.38	0.64	40.37	0.15
3	11.15	15.42	−0.52	20.70	0.41	13.36	9.93	−0.08	11.57	0.86	2.22	11.80	0.99	54.50	0.19
4	10.95	15.80	−0.71	20.94	0.39	13.52	10.21	−0.23	16.48	0.86	2.57	12.05	1.34	53.11	0.21
5	11.13	15.62	−0.45	23.02	0.41	11.40	10.10	−0.46	19.56	0.66	0.28	11.92	0.61	57.78	0.02
6	11.89	15.14	−0.45	16.96	0.47	13.36	9.81	−0.33	15.51	0.87	1.47	11.53	0.66	42.20	0.13
7	12.55	14.89	−0.64	22.73	0.52	13.91	9.75	−0.35	14.88	0.94	1.36	11.26	1.10	61.08	0.12
8	12.98	15.82	−0.60	30.16	0.52	14.09	10.26	−0.50	33.44	0.91	1.10	12.04	0.90	72.09	0.09
9	13.90	15.99	−0.68	20.87	0.57	15.50	10.48	−0.52	17.94	1.02	1.60	12.07	1.07	53.79	0.13
High	15.01	17.78	−0.46	15.88	0.58	18.58	11.73	−0.32	17.48	1.18	3.57	13.35	0.68	39.36	0.27
Panel C: Momentum-sorted portfolios															
Low	2.59	24.93	0.41	25.70	−0.09	18.75	15.86	1.09	58.91	0.88	16.15	19.20	−0.33	45.63	0.84
2	8.03	20.35	0.15	22.17	0.16	16.58	13.27	−0.06	40.66	0.89	8.54	15.41	−0.48	45.21	0.55
3	10.29	17.59	0.04	17.98	0.31	15.86	11.68	0.32	31.61	0.95	5.58	13.13	0.02	37.98	0.42
4	10.22	16.61	−0.15	18.81	0.33	13.85	10.85	0.41	19.86	0.84	3.63	12.57	0.50	46.24	0.29
5	10.09	15.89	0.01	21.65	0.33	11.90	10.48	0.17	26.00	0.68	1.81	11.95	−0.01	52.40	0.15
6	10.99	15.35	−0.58	26.71	0.40	12.27	10.02	−0.35	19.35	0.75	1.28	11.63	1.00	70.35	0.11
7	10.92	15.19	−0.55	22.88	0.40	11.88	9.79	−0.24	12.56	0.73	0.96	11.63	0.97	60.36	0.08
8	13.57	15.49	−0.73	21.21	0.57	13.26	10.12	−0.37	11.34	0.84	−0.32	11.74	1.30	57.95	−0.03
9	12.81	16.47	−0.54	15.10	0.49	13.40	10.41	−0.35	10.08	0.83	0.59	12.77	0.83	37.25	0.05
High	17.85	20.37	−0.50	12.74	0.64	19.80	12.65	−0.59	9.59	1.19	1.95	15.97	0.59	29.82	0.12

continued

Table 2.1 Continued

Portfolio	BH Portfolios					MA Portfolios					MAP Portfolios				
	μ	σ	s	k	SR	μ	σ	s	k	SR	μ	σ	s	k	SR
Panel D: Short-term reversal sorted portfolios															
Low	41.02	25.55	0.87	29.76	1.42	35.19	15.11	0.12	23.04	2.01	−5.84	20.64	−1.93	63.99	−0.28
2	18.83	20.36	0.33	27.47	0.69	18.91	11.68	0.08	23.42	1.21	0.09	16.68	−0.75	55.55	0.01
3	14.32	17.97	0.06	26.89	0.53	15.38	10.57	−0.35	23.91	1.00	1.05	14.54	−0.39	56.25	0.07
4	11.57	16.65	−0.38	23.86	0.41	14.45	10.09	−0.03	19.20	0.96	2.88	13.25	0.62	53.05	0.22
5	11.73	15.81	−0.47	22.77	0.44	13.93	9.71	0.03	13.32	0.94	2.20	12.47	0.84	53.78	0.18
6	12.78	15.57	−0.45	22.73	0.51	14.80	10.08	−0.05	12.50	0.99	2.02	11.86	0.83	60.79	0.17
7	9.63	15.45	−0.55	19.99	0.31	11.55	10.22	−0.29	15.10	0.66	1.92	11.59	1.00	53.96	0.17
8	7.93	15.66	−0.42	16.68	0.20	11.00	10.79	−0.24	19.73	0.58	3.08	11.35	0.82	44.33	0.27
9	6.01	16.52	−0.44	15.75	0.07	8.44	11.69	−0.72	23.32	0.31	2.43	11.67	0.49	39.82	0.21
High	−9.59	19.08	−0.55	11.88	−0.75	1.16	12.80	−1.11	25.80	−0.28	10.74	14.17	0.68	22.11	0.76
Panel E: Long-term reversal sorted portfolios															
Low	14.73	20.02	−0.38	14.61	0.50	21.32	13.05	0.21	12.88	1.27	6.60	15.16	0.85	37.08	0.44
2	13.72	17.10	−0.61	18.75	0.52	15.95	11.14	−0.35	14.60	1.00	2.23	12.97	1.02	48.52	0.17
3	13.09	15.94	−0.58	18.32	0.52	14.03	10.61	−0.18	12.21	0.87	0.94	11.90	1.11	51.17	0.08
4	12.96	15.66	−0.79	25.23	0.52	12.99	10.35	−0.25	13.65	0.79	0.03	11.75	1.56	71.08	0.00
5	11.63	15.50	−0.71	22.89	0.44	12.74	10.25	−0.12	17.72	0.78	1.11	11.63	1.47	61.37	0.10
6	12.91	14.73	−0.36	21.45	0.55	13.24	9.75	−0.20	12.63	0.87	0.33	11.05	0.56	60.08	0.03
7	12.10	15.03	−0.39	17.81	0.49	12.65	9.96	−0.17	12.95	0.79	0.55	11.26	0.68	48.59	0.05
8	11.03	15.60	−0.46	19.22	0.40	12.21	9.94	−0.12	13.34	0.75	1.17	12.02	0.83	48.19	0.10
9	9.75	16.87	−0.19	18.39	0.29	12.05	10.57	−0.19	13.88	0.69	2.30	13.15	0.21	44.06	0.17
High	10.66	20.33	−0.16	14.88	0.29	14.55	12.41	−0.33	14.03	0.79	3.89	16.11	0.09	32.85	0.24

						Panel F: Volatility-sorted portfolios									
High	40.10	20.27	0.10	14.40	1.74	52.94	14.06	1.34	18.57	3.43	12.84	14.43	0.47	38.31	0.89
2	17.80	19.30	−0.31	14.52	0.67	31.39	12.24	0.56	17.27	2.17	13.59	14.83	0.77	33.11	0.92
3	15.22	18.25	−0.38	16.39	0.57	27.28	11.49	0.05	20.62	1.96	12.06	14.11	0.64	36.52	0.85
4	15.74	16.89	−0.47	18.77	0.65	24.92	10.71	−0.18	23.81	1.88	9.18	13.01	0.72	42.05	0.71
5	15.68	15.52	−0.55	20.47	0.70	23.94	9.86	−0.30	25.40	1.94	8.27	11.93	0.80	46.31	0.69
6	14.82	14.38	−0.67	24.45	0.70	21.45	9.15	−0.57	28.55	1.82	6.63	11.06	0.91	56.17	0.60
7	14.40	12.78	−0.66	28.32	0.75	20.75	8.07	−0.39	20.56	1.98	6.35	9.87	0.97	69.92	0.64
8	13.68	11.15	−0.78	36.32	0.80	20.18	6.92	0.07	11.71	2.23	6.50	8.71	1.40	92.46	0.75
9	12.53	9.27	−0.58	54.72	0.84	18.90	5.62	0.27	11.34	2.51	6.37	7.33	1.02	135.67	0.87
Low	10.81	6.86	−0.47	71.59	0.88	16.86	4.12	0.65	15.21	2.94	6.05	5.44	0.90	175.78	1.11
						Panel G: Industry-sorted portfolios									
NoDur	12.80	13.63	−0.62	22.17	0.59	15.65	9.04	−0.16	12.18	1.20	2.85	10.19	1.22	63.10	0.28
Durbl	10.57	20.63	−0.23	11.77	0.28	13.16	13.75	0.13	15.08	0.61	2.59	15.38	0.56	28.39	0.17
Manuf	11.07	16.28	−0.68	21.25	0.39	14.83	10.47	−0.27	14.56	0.96	3.75	12.46	1.23	54.42	0.30
Enrgy	13.49	19.99	−0.20	19.86	0.44	12.56	13.45	−0.27	14.34	0.58	−0.93	14.80	0.18	56.38	−0.06
HiTec	11.37	22.42	0.02	12.00	0.29	15.53	14.14	0.16	10.66	0.76	4.16	17.39	−0.05	28.48	0.24
Telcm	10.94	17.36	−0.03	17.12	0.36	10.99	11.49	0.10	13.87	0.54	0.05	13.01	0.04	45.81	0.00
Shops	12.23	16.54	−0.28	14.21	0.45	15.87	10.84	0.03	12.64	1.02	3.64	12.49	0.54	36.47	0.29
Hlth	12.52	16.83	−0.40	14.37	0.46	14.54	11.10	−0.13	11.27	0.88	2.03	12.65	0.74	38.24	0.16
Utils	10.18	12.93	−0.03	28.88	0.42	14.01	8.33	−0.25	16.46	1.11	3.84	9.88	−0.25	76.60	0.39
Other	11.15	17.73	−0.26	18.07	0.36	15.61	11.19	−0.14	29.11	0.97	4.47	13.75	0.36	37.18	0.32

Furthermore, there appear to be some substantial cross-sectional differences related to the size effect (Panel A), the value premium (Panel B), as well as momentum premia (Panel C).

From the evidence presented in Table 2.1 it appears that portfolios with higher standard deviations tend to experience higher average improvements between the buy-and-hold and the MA strategy performance or $\Delta\mu$. A more formal way to test this is through a cross-sectional regression which is presented next:

$$\underset{}{\Delta\mu} = \underset{(1.08)}{-2.19} + \underset{(0.06)}{0.35}\sigma, \tag{2.5}$$

where the cross-sectional $R^2 = 0.5834$ and the standard errors are corrected for heteroscedasticity and reported in parentheses. The slope on σ is highly statistically and economically significant suggesting that on average $\Delta\mu$ increases by 3.5% on an annualized basis when the portfolio return volatility increases by 10%.[4]

The MA strategy clearly performs very well compared to the BH strategy. The next section investigates more formally the reasons for this performance in the traditional empirical asset pricing framework of factor models and abnormal returns.

2.3 Abnormal Returns

The first asset pricing model I consider is the CAPM of Sharpe (1964):

$$\text{MAP}_{jt,L} = \alpha_j + \beta_{j,m} r_{mkt,t} + \epsilon_{jt}, \quad j = 1, \ldots, N, \tag{2.6}$$

where $r_{mkt,t}$ is the excess return on the market portfolio at the end of month t. The panel on the left in Table 2.2 presents the annualized alphas in percent and the market betas of the MAP excess returns for several sets of portfolios as well as the High minus Low cross-sectional difference.The CAPM alphas of portfolios sorted by market size range between 2.5% for the largest and 16.5% per year for the smallest size, respectively. The alphas of book-to-market portfolios range between 3.6% for decile 5 and 7.1% for the High decile, respectively. Portfolios sorted on momentum deliver an even wider range of CAPM alphas (2.9% for decile 8 to 21.1% for the Low decile). Portfolios sorted on past short-term reversal exhibit monotonically increasing alphas from a −0.8% for the Low decile to 14.7% for the High decile. The abnormal returns for portfolios sorted on past long-term reversal are less monotonic across the deciles ranging from a low of 3.2% for decile 4 to 10.7% for the Low decile. The abnormal returns for portfolio sorted on total volatility are almost monotonic yielding alphas of between 7.1% for the low volatility decile portfolio and 17.5% for decile 2, the second-highest volatility portfolio. Finally, the CAPM alphas for portfolios sorted by industry yield a low and statistically insignificant abnormal return of 2.4% for the Enrgy portfolio and a relatively high and statistically significant 8.9% for the HiTec portfolio.

The next asset pricing model I use to adjust for the risk of the MAP returns is the Fama and French (1992) three-factor model:

$$\mathrm{MAP}_{jt,L} = \alpha_j + \beta_{j,m} r_{mkt,t} + \beta_{j,s} r_{smb,t} + \beta_{j,h} r_{hml,t} + \epsilon_{jt}, \quad j = 1, \ldots, N, \quad (2.7)$$

where $r_{mkt,t}$ is the excess market return at the end of month t, $r_{smb,t}$ is the return on the SMB factor at the end of month t, and $r_{hml,t}$ is the return on the HML factor at the end of month t. The middle panel of Table 2.2 reports the empirical results. The Fama-French alphas are largely similar to the CAPM alphas with some of the previous results for the High minus Low portfolio becoming statistically significant (size deciles in Panel A), some becoming statistically insignificant with the rest unchanged qualitatively. The sign and magnitude of the Fama-French market betas is largely the same as the CAPM market betas, namely, significantly negative though low in absolute value, suggesting that the MAP returns can be used to hedge exposure to the underlying portfolios. The small minus big (SMB) betas are either negative and significant or insignificant. Similarly, the high minus low (HML) betas are mostly negative and significant though of smaller absolute values than the market betas. In terms of factor attribution, these results suggest that the MAP returns have negative exposure to the market factor, a somewhat positive exposure on larger stocks with an emphasis on growth stocks over value stocks.

The final asset pricing model I consider in this section is the four-factor Carhart (1997) model:

$$\begin{aligned} \mathrm{MAP}_{jt,L} = \alpha_j &+ \beta_{j,m} r_{mkt,t} + \beta_{j,s} r_{smb,t} + \beta_{j,h} r_{hml,t} \\ &+ \beta_{j,u} r_{umd,t} + \epsilon_{jt}, \quad j = 1, \ldots, N, \end{aligned} \quad (2.8)$$

where all variables are as defined before and $r_{umd,t}$ is the return of the UMD factor, or the momentum factor, at the end of month t. The panel on the right in Table 2.2 presents the results for the four-factor model adjustment to the MAP returns. First, note that the vast majority of the risk-adjusted alphas are lower than the CAPM and the Fama-French alphas. However, they are all still quite substantial economically and are still highly statistically significant. The factor loadings on the market portfolio, SMB, and HML are largely unchanged while the loadings on the UMD factor are mostly negative and statistically significant (with the exception of decile 1 in some panels). This is in line with the CAPM factor loadings reported previously and further demonstrates the contrarian nature of the MAP returns.

Reading Table 2.2 across, we notice that the adjusted R^2 improves when we consider in turn the Fama-French three-factor model and the four-factor Carhart model. This improvement is uniform across the different sets of portfolios I consider and suggests that all four factors have a role to play in driving the performance of the MAP returns. Nevertheless, the average adjusted R^2 values suggest that only around half of the return variation can be explained and accounted for by market, size, value, and momentum. This leaves a large portion of return variation that cannot be accounted for.

2.4 Discussion

The large values of the risk-adjusted abnormal returns presented in the previous subsection demonstrate the profitability of the MA switching strategy. This raises the question as to what are the drivers of the performance of the MA strategy. So far the evidence points toward a strategy that is contrarian, with a focus on large-cap growth stocks and short the market. However, the goodness-of-fit statistics so far indicate this is at most only half the story. A more fundamental question that arises is how can this strategy survive in competitive financial markets. A few potential reasons seem plausible.

First, there is ample evidence that stock returns are predictable at various frequencies at least to a certain degree. This level of predictability is not perfect but is sufficient to improve forecasts of future stock returns when stock return predictability is ignored. Some of the early evidence presented in Fama and Schwert (1977) and Campbell (1987) as well as more recent work by Cochrane (2008) clearly demonstrates that stock return predictability is an important feature that investors should ignore at their own peril.

Evidence regarding the performance of the MA technical indicator is present in Brock, Lakonishok and LeBaron (1992) in the context of predicting future moments of the Dow Jones Industrial Average. Lo, Mamaysky and Wang (2000) provide further evidence using a wide range of technical indicators with wide popularity among traders showing that this adds value even at the individual stock level over and above the

Table 2.2 Factor regressions results.

This table reports alphas, betas, and adjusted R^2 of the regressions of the MAP excess returns on the market factor, the Fama-French three-factors, and the Carhart four-factors using portfolios sorted by various characteristics. The alphas are annualized and in percent. The sample period covers January 4, 1960 until December 31, 2013 with daily value-weighted portfolio returns. The length of the MA window is 20 days. Newey and West (1987) standard errors with five lags are used in reporting statistical significance of a two-sided null hypothesis at the 1%, 5%, and 10% level given by a ***, a **, and a *, respectively.

Portfolio	α	β_m	$\bar{R}^2$	α	β_m	β_s	β_b	$\bar{R}^2$	α	β_m	β_s	β_b	β_u	$\bar{R}^2$
	CAPM			Fama-French					Carhart					
Panel A: Size-sorted portfolios														
Low	16.522***	−0.435***	0.404	18.194***	−0.492***	−0.526***	−0.133***	0.555	19.115***	−0.504***	−0.524***	−0.164***	−0.086***	0.562
2	13.493***	−0.561***	0.453	15.394***	−0.626***	−0.604***	−0.149***	0.586	16.343***	−0.638***	−0.601***	−0.181***	−0.089***	0.591
3	12.074***	−0.575***	0.482	13.776***	−0.633***	−0.559***	−0.127***	0.598	14.648***	−0.645***	−0.557***	−0.157***	−0.081***	0.602
4	11.379***	−0.572***	0.494	12.753***	−0.620***	−0.505***	−0.085***	0.590	13.750***	−0.633***	−0.502***	−0.118***	−0.093***	0.596
5	9.860***	−0.576***	0.511	10.954***	−0.616***	−0.453***	−0.050***	0.589	11.755***	−0.626***	−0.451***	−0.077***	−0.075***	0.593
6	9.619***	−0.554***	0.528	10.313***	−0.580***	−0.319***	−0.021**	0.572	11.166***	−0.591***	−0.317***	−0.050***	−0.080***	0.577
7	9.380***	−0.566***	0.540	10.071***	−0.590***	−0.261***	−0.040***	0.568	10.980***	−0.602***	−0.258***	−0.071***	−0.085***	0.574
8	8.026***	−0.583***	0.560	8.643***	−0.604***	−0.184***	−0.053***	0.574	9.500***	−0.615***	−0.181***	−0.081***	−0.080***	0.578
9	5.702***	−0.591***	0.583	6.168***	−0.605***	−0.066***	−0.065***	0.586	7.012***	−0.615***	−0.063***	−0.093***	−0.079***	0.591
High	2.478**	−0.612***	0.589	1.896*	−0.594***	0.143***	0.060***	0.598	2.400**	−0.600***	0.145***	0.043***	−0.047***	0.600

continued

Table 2.2 Continued

Portfolio	α	β_m	$\bar{R}^2$	α	β_m	β_s	β_h	$\bar{R}^2$	α	β_m	β_s	β_h	β_u	$\bar{R}^2$
	CAPM			Fama-French					Carhart					
Panel B: Book-to-market sorted portfolios														
Low	6.934***	−0.663***	0.564	5.149***	−0.615***	0.037***	0.321***	0.592	5.592***	−0.621***	0.038***	0.306***	−0.041***	0.593
2	5.519***	−0.600***	0.565	4.924***	−0.584***	0.021**	0.104***	0.568	5.684***	−0.593***	0.023**	0.079***	−0.071***	0.572
3	5.613***	−0.562***	0.546	5.384***	−0.556***	0.021**	0.036***	0.547	6.441***	−0.569***	0.023***	0.000	−0.099***	0.554
4	6.035***	−0.574***	0.546	6.742***	−0.593***	−0.009	−0.129***	0.552	7.469***	−0.602***	−0.007	−0.153***	−0.068***	0.555
5	3.619***	−0.554***	0.520	4.597***	−0.580***	−0.013	−0.178***	0.531	5.192***	−0.587***	−0.011	−0.198***	−0.056***	0.533
6	4.696***	−0.535***	0.518	5.784***	−0.565***	−0.074***	−0.178***	0.531	6.379***	−0.573***	−0.072***	−0.198***	−0.056***	0.534
7	4.456***	−0.513***	0.500	5.862***	−0.551***	−0.035***	−0.251***	0.526	6.727***	−0.562***	−0.033***	−0.280***	−0.081***	0.531
8	4.309***	−0.531***	0.469	6.752***	−0.598***	−0.076***	−0.430***	0.535	7.112***	−0.602***	−0.075***	−0.443***	−0.034***	0.536
9	4.858***	−0.540***	0.482	6.990***	−0.598***	−0.067***	−0.375***	0.532	8.010***	−0.611***	−0.064***	−0.409***	−0.095***	0.539
High	7.085***	−0.582***	0.458	9.731***	−0.656***	−0.188***	−0.430***	0.518	10.696***	−0.669***	−0.186***	−0.462***	−0.090***	0.523
Panel C: Momentum sorted portfolios														
Low	21.170***	−0.831***	0.451	22.735***	−0.879***	−0.278***	−0.198***	0.467	18.075***	−0.819***	−0.290***	−0.042**	0.435***	0.523
2	12.757***	−0.698***	0.494	13.644***	−0.723***	−0.085***	−0.137***	0.499	10.335***	−0.680***	−0.094***	−0.026**	0.309***	0.542
3	9.228***	−0.605***	0.511	9.938***	−0.624***	−0.005	−0.131***	0.516	7.911***	−0.598***	−0.011	−0.063***	0.189***	0.538
4	7.200***	−0.592***	0.533	7.819***	−0.608***	0.017*	−0.121***	0.538	7.053***	−0.598***	0.015	−0.096***	0.072***	0.542
5	5.187***	−0.560***	0.528	5.581***	−0.570***	−0.003	−0.073***	0.530	5.562***	−0.570***	−0.003	−0.072***	0.002	0.530
6	4.549***	−0.541***	0.521	4.834***	−0.548***	0.022**	−0.061***	0.523	5.491***	−0.556***	0.024***	−0.083***	−0.061***	0.526
7	4.259***	−0.547***	0.532	4.546***	−0.554***	0.008	−0.056***	0.533	5.951***	−0.572***	0.012	−0.103***	−0.131***	0.547
8	2.952***	−0.541***	0.512	3.280***	−0.550***	0.015	−0.066***	0.514	5.468***	−0.578***	0.021**	−0.140***	−0.204***	0.547
9	4.194***	−0.596***	0.525	4.402***	−0.604***	−0.081***	−0.011	0.528	7.202***	−0.640***	−0.074***	−0.105***	−0.261***	0.573
High	6.229***	−0.708***	0.474	5.725***	−0.701***	−0.237***	0.175***	0.495	10.109***	−0.757***	−0.226***	0.028**	−0.409***	0.566

Panel D: Short-term reversal sorted portfolios														
Low	−0.811	−0.832***	0.391	−0.282	−0.852***	−0.237***	−0.018	0.399	−1.179	−0.840***	−0.239***	0.012	0.084***	0.401
2	4.654***	−0.756***	0.495	4.983***	−0.768***	−0.098***	−0.028**	0.497	5.008***	−0.768***	−0.098***	−0.029**	−0.002	0.497
3	5.092***	−0.669***	0.510	5.389***	−0.678***	−0.053***	−0.037***	0.511	6.052***	−0.687***	−0.051***	−0.060***	−0.062***	0.513
4	6.641***	−0.624***	0.534	6.844***	−0.629***	−0.011	−0.034***	0.534	7.695***	−0.640***	−0.009	−0.063***	−0.079***	0.538
5	5.803***	−0.597***	0.551	6.076***	−0.604***	−0.014	−0.046***	0.552	6.825***	−0.614***	−0.013	−0.071***	−0.070***	0.555
6	5.406***	−0.561***	0.539	5.689***	−0.569***	−0.008	−0.050***	0.540	6.499***	−0.579***	−0.006	−0.077***	−0.076***	0.544
7	5.288***	−0.557***	0.557	5.593***	−0.566***	−0.015*	−0.052***	0.558	6.272***	−0.575***	−0.013	−0.075***	−0.063***	0.562
8	6.339***	−0.541***	0.546	6.551***	−0.547***	−0.020**	−0.033***	0.547	7.538***	−0.559***	−0.018**	−0.066***	−0.092***	0.554
9	5.731***	−0.547***	0.529	6.163***	−0.559***	−0.049***	−0.064***	0.531	6.657***	−0.566***	−0.047***	−0.080***	−0.046***	0.533
High	14.735***	−0.661***	0.524	15.216***	−0.678***	−0.186***	−0.026**	0.535	15.235***	−0.679***	−0.186***	−0.027**	−0.002	0.535
Panel E: Long-term reversal sorted portfolios														
Low	10.695***	−0.679***	0.483	12.441***	−0.733***	−0.340***	−0.210***	0.520	13.931***	−0.752***	−0.336***	−0.260***	−0.139***	0.529
2	5.836***	−0.597***	0.510	7.062***	−0.633***	−0.153***	−0.177***	0.526	8.469***	−0.651***	−0.149***	−0.224***	−0.131***	0.537
3	4.226***	−0.544***	0.504	5.258***	−0.573***	−0.063***	−0.171***	0.515	6.478***	−0.589***	−0.060***	−0.212***	−0.114***	0.525
4	3.271***	−0.537***	0.503	4.394***	−0.568***	−0.031***	−0.199***	0.518	5.168***	−0.578***	−0.029***	−0.225***	−0.072***	0.522
5	4.333***	−0.534***	0.507	5.414***	−0.562***	−0.008	−0.199***	0.522	6.194***	−0.572***	−0.006	−0.225***	−0.073***	0.526
6	3.402***	−0.509***	0.510	3.880***	−0.520***	0.045***	−0.105***	0.516	4.798***	−0.532***	0.047***	−0.135***	−0.086***	0.523
7	3.700***	−0.521***	0.516	3.927***	−0.526***	0.041***	−0.056***	0.518	4.782***	−0.537***	0.043***	−0.085***	−0.080***	0.524
8	4.611***	−0.569***	0.540	4.576***	−0.567***	0.054***	−0.012	0.541	5.654***	−0.581***	0.057***	−0.048***	−0.101***	0.549
9	6.052***	−0.621***	0.538	5.775***	−0.613***	0.051***	0.034***	0.539	6.669***	−0.624***	0.054***	0.004	−0.083***	0.543
High	8.508***	−0.765***	0.543	7.209***	−0.732***	−0.070***	0.267***	0.559	7.460***	−0.736***	−0.070***	0.258***	−0.023***	0.559

continued

Table 2.2 Continued

Portfolio	α	β_m	$\bar{R}^2$	α	β_m	β_s	β_h	$\bar{R}^2$	α	β_m	β_s	β_h	β_u	$\bar{R}^2$
	CAPM			Fama-French					Carhart					
						Panel F: Volatility sorted portfolios								
High	16.170***	−0.552***	0.353	17.937***	−0.612***	−0.535***	−0.148***	0.437	18.588***	−0.620***	−0.534***	−0.169***	−0.061***	0.439
2	17.511***	−0.650***	0.462	19.221***	−0.707***	−0.506***	−0.147***	0.534	19.731***	−0.713***	−0.504***	−0.164***	−0.048***	0.535
3	15.864***	−0.630***	0.480	17.595***	−0.687***	−0.455***	−0.168***	0.546	18.298***	−0.696***	−0.453***	−0.192***	−0.066***	0.549
4	12.705***	−0.584***	0.485	14.421***	−0.638***	−0.383***	−0.190***	0.544	15.010***	−0.646***	−0.381***	−0.210***	−0.055***	0.546
5	11.499***	−0.535***	0.485	13.150***	−0.586***	−0.310***	−0.203***	0.536	13.746***	−0.594***	−0.309***	−0.223***	−0.056***	0.538
6	9.645***	−0.499***	0.490	11.140***	−0.544***	−0.242***	−0.197***	0.531	11.708***	−0.551***	−0.241***	−0.216***	−0.053***	0.533
7	9.021***	−0.443***	0.484	10.138***	−0.476***	−0.168***	−0.151***	0.511	10.809***	−0.485***	−0.166***	−0.174***	−0.063***	0.515
8	8.817***	−0.384***	0.469	9.727***	−0.411***	−0.114***	−0.131***	0.489	10.398***	−0.420***	−0.112***	−0.153***	−0.063***	0.494
9	8.204***	−0.304***	0.414	8.980***	−0.326***	−0.075***	−0.120***	0.432	9.514***	−0.333***	−0.073***	−0.137***	−0.050***	0.437
Low	7.095***	−0.173***	0.244	7.380***	−0.181***	−0.024***	−0.045***	0.248	7.891***	−0.188***	−0.022***	−0.062***	−0.048***	0.257
						Panel G: Industry-sorted portfolios								
NoDur	5.467***	−0.434***	0.436	5.357***	−0.429***	0.063***	−0.001	0.438	6.509***	−0.444***	0.066***	−0.039***	−0.108***	0.450
Durbl	6.670***	−0.676***	0.465	7.808***	−0.707***	−0.041***	−0.199***	0.474	8.320***	−0.714***	−0.039***	−0.216***	−0.048***	0.475
Manuf	7.326***	−0.592***	0.543	7.729***	−0.603***	−0.023**	−0.067***	0.545	8.777***	−0.617***	−0.020**	−0.102***	−0.098***	0.552
Enrgy	2.422	−0.554***	0.338	3.058*	−0.568***	0.117***	−0.159***	0.349	4.557***	−0.588***	0.121***	−0.209***	−0.140***	0.358
HiTec	8.895***	−0.785***	0.491	6.444***	−0.721***	−0.047***	0.474***	0.530	6.338***	−0.720***	−0.047***	0.477***	0.010	0.530
Telcm	3.301**	−0.539***	0.413	3.699***	−0.547***	0.110***	−0.112***	0.422	3.787***	−0.548***	0.110***	−0.115***	−0.008	0.422
Shops	6.968***	−0.551***	0.468	6.803***	−0.547***	−0.034***	0.042***	0.469	7.585***	−0.557***	−0.032***	0.016	−0.073***	0.473
Hlth	5.195***	−0.525***	0.414	4.320***	−0.499***	0.090***	0.133***	0.422	5.429***	−0.514***	0.093***	0.096***	−0.104***	0.430
Utils	5.940***	−0.349***	0.300	7.031***	−0.376***	0.065***	−0.226***	0.331	7.880***	−0.387***	0.067***	−0.254***	−0.079***	0.338
Other	8.385***	−0.649***	0.537	9.958***	−0.694***	−0.139***	−0.246***	0.557	10.114***	−0.696***	−0.138***	−0.252***	−0.015*	0.557

performance of a stock index. More recently, Neely, Rapach, Tu and Zhou (2010) provide evidence in favor of the usefulness of technical analysis in forecasting the stock market risk premium.

Second, early work on the performance of filter rules by Fama and Blume (1966) and Jensen and Benington (1970) concluded that such rules were dominated by buy-and-hold strategies especially after transaction costs. Malkiel (1996) makes a forceful and memorable point against technical indicators: *"Obviously, I'm biased against the chartist. This is not only a personal predilection but a professional one as well. Technical analysis is anathema to the academic world. We love to pick on it. Our bullying tactics are prompted by two considerations: (1) after paying transaction costs, the method does not do better than a buy-and-hold strategy for investors, and (2) it's easy to pick on. And while it may seem a bit unfair to pick on such a sorry target, just remember: It's your money we are trying to save."* In a follow-up on Brock et al. (1992), Bessembinder and Chan (1998) attribute the forecasting power of technical analysis to measurement errors arising from non-synchronous trading. Ready (2002) goes even further and claims the results in Brock et al. (1992) are spurious and due to data snooping. Formal tests using White's Reality Check are conducted in Sullivan, Timmerman and White (1999) and confirm that Brock et al.'s (1992) results are robust to data snooping and perform even better out of sample though there is evidence of time variation in performance across subperiods. A more recent study using White's Reality Check and Hansen's SPA test is Hsu and Kuan (2005) who find evidence of profitability of technical analysis using relatively "young" markets like the NASDAQ Composite index and the Russell 2000, both in-sample and out-of-sample.

Furthermore, Treynor and Ferguson (1985) make a strong case in favor of investor's learning and Bayesian updating conditional on new information received rationally combining past prices that can result in abnormal profitability. Sweeney (1988) revisits Fama and Blume (1966) and finds that fulter rules can be profitable to floor traders in the 1970–1982 time period. Neftci (1991) presents a formal analysis of Wiener–Kolmogorov prediction theory which provides optimal linear forecasts. He concludes that if the underlying price processes are nonlinear in nature then technical analysis rules might capture some useful information that is ignored by the linear prediction rules. More involved and inherently nonlinear rules are investigated in the context of foreign currency exchange rates by Neely, Weller and Dittmar (1997) using a genetic programming approach. Gencay (1998) goes even further in using nonlinear predictors based on simple MA rules on the Dow Jones Industrial Average over a long time period between 1897 and 1988. In a similar vein, Allen and Karjalainen (1999) use genetic algorithm to search for functions of past prices that can outperform a simple buy-and-hold strategy and report negative excess returns for most of the strategies they consider.

Thirdly, it is entirely possible that market prices of financial assets can persistently deviate from fundamental values. Those fundamental values themselves are subject to incomplete information and, perhaps, imperfect understanding of valuation tools as well as dispersion of beliefs and objective and behavioral biases across the pool of traders and investors who regularly interact in financial markets. When investors' information is incomplete and they learn continuously over time the true fundamental value, Zhu and Zhou (2009) show theoretically that the MA price is a useful state variable that aids in investors' learning and improves their well-being and utility.

Behavioral and cognitive biases have been proposed in Daniel, Hirshleifer and Subrahmanyam (1998) and Hong and Stein (1999), among others, as a potential driver of both price under- and over-reaction in conjunction with the observed price continuation of stock prices. An alternative explanation for price continuation was proposed in Zhang (2006). He argues that investors suboptimally underweight newly arriving public information leading to a persistent deviation of the market price from the fundamental intrinsic value.

Note also that despite the apparent similarity of the MA switching strategy to the momentum strategy, the four-factor alphas presented in the previous subsection are still mostly significant and of high magnitudes. More appropriately, the payoff of the MA strategy is similar to the payoffs of an at-the-money protective put option that is continuously reloaded as the MA window moves forward in time. Hence, perhaps it should be less surprising that it reduces volatility compared to the buy-and-hold portfolio and improves the mean return (due to the positive convexity of the combined position).

2.5 Explanation

Before making an attempt at explaining the reasons for the profitability of the MA strategies performance, it is useful to inspect a scatter plot of the MA strategy returns versus the underlying BH strategy returns for the same portfolio. Figure 2.1 presents the scatter plot for the first decile of the market-capitalization sorted deciles. Figure 2.2 plots the scatter plots for the portfolios sorted on book-to-market. Figure 2.3 presents the scatter

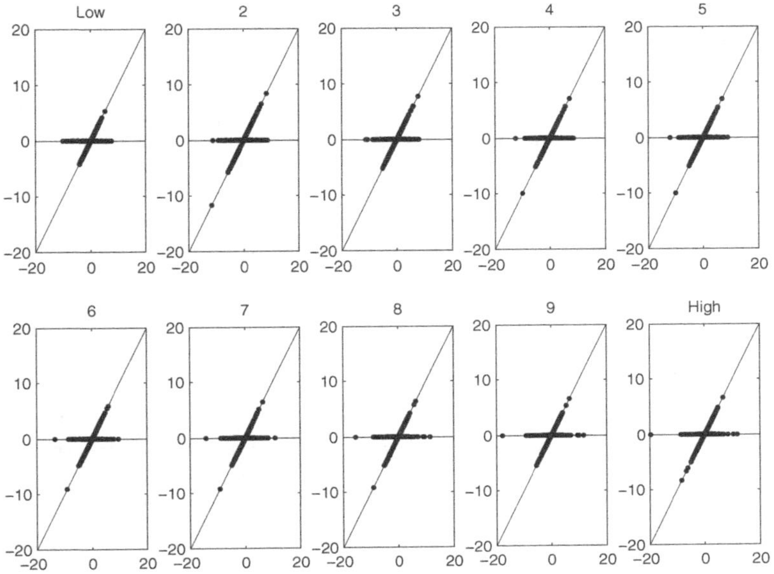

Figure 2.1 Scatter plot of BH versus MA returns: Size decile portfolios.

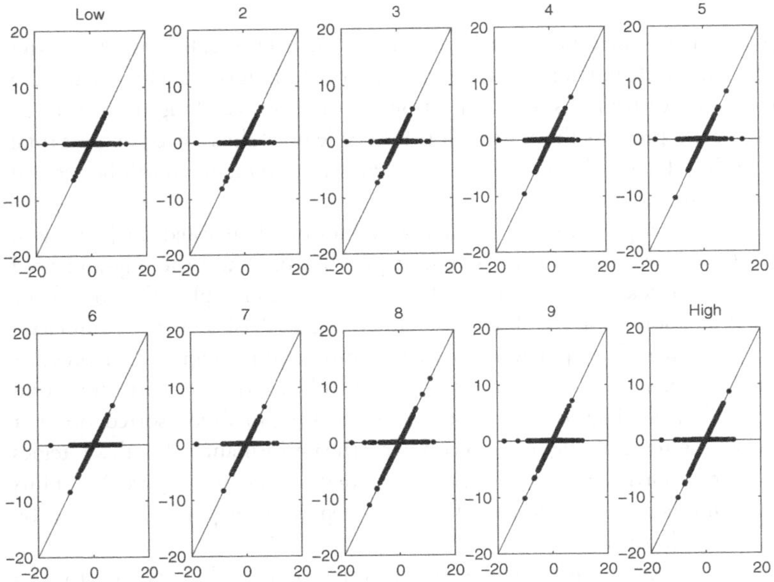

Figure 2.2 Scatter plot of BH versus MA returns: Book-to-market decile portfolios.

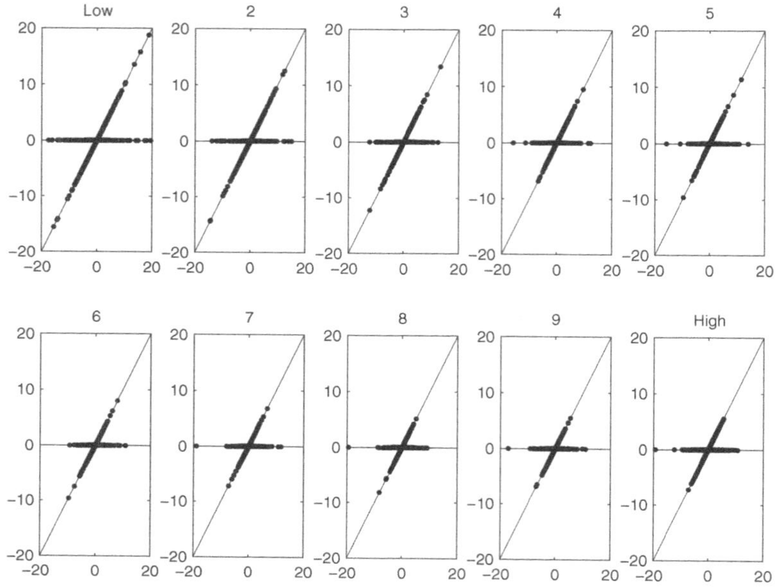

Figure 2.3 Scatter plot of BH versus MA returns: Momentum decile portfolios.

plots for the portfolios sorted on momentum. Figure 2.4 shows the scatter plots for decile portfolios sorted on past short-term reversal while Figure 2.5 plots the scatter plots for the portfolios sorted on past long-term reversal. Figure 2.6 presents the scatter plots for decile portfolios sorted on past total volatility. Finally, Figure 2.7 presents the scatter plots for portfolios sorted by industry classification.

Figure 2.8 presents the cumulative values of $1 invested on January 4, 1960, onward for both the MA strategy (thin line) and the buy-and-hold strategy (thick line). Figure 2.9 presents the same plot for the decile portfolios sorted on past book-to-market. Figure 2.10 plots the cumulative performance of the portfolios sorted on momentum. Figure 2.11 presents the performance of both strategies for portfolios sorted on past short-term reversal while Figure 2.12 plots the same for portfolios sorted on past long-term reversal. Figure 2.13 compares the performance of both strategies for decile portfolios sorted on past total volatility. Finally, Figure 2.14 plots the cumulative total returns of both strategies using portfolios sorted by industry classification.

The strategy clearly triggers false-positive signals where we are told to stay invested or switch into the underlying asset with a subsequent negative return (negative quadrant of returns in the figure). Similarly, there are a

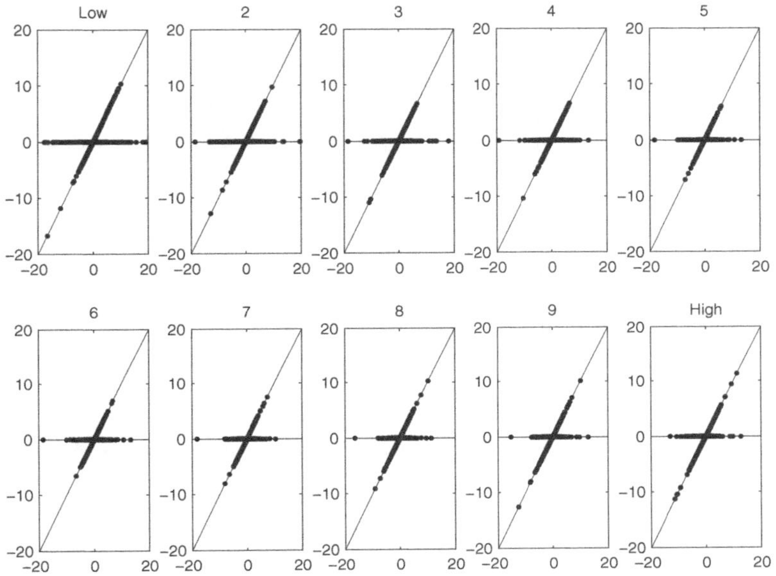

Figure 2.4 Scatter plot of BH versus MA returns: Short-term reversal decile portfolios.

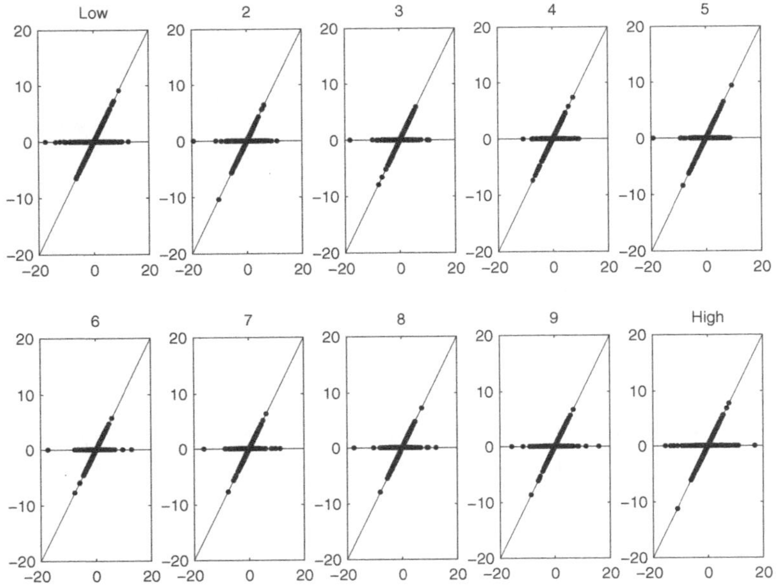

Figure 2.5 Scatter plot of BH versus MA returns: Long-term reversal decile portfolios.

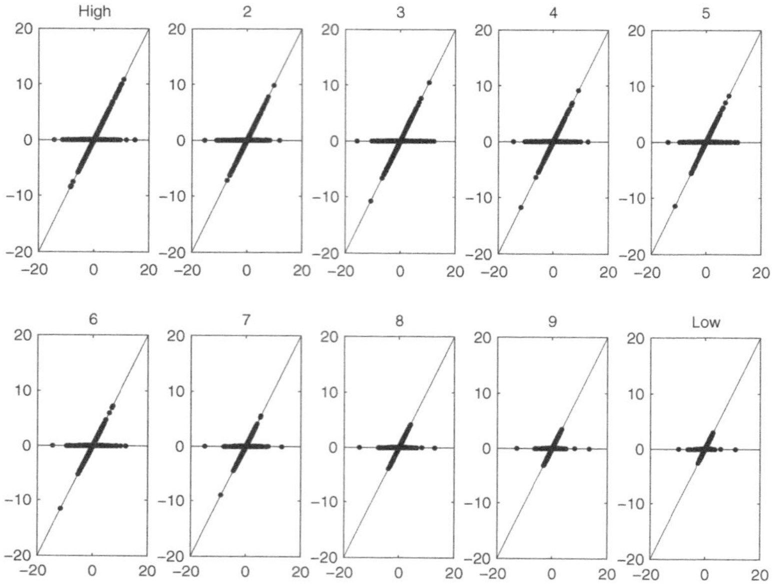

Figure 2.6 Scatter plot of BH versus MA returns: Volatility decile portfolios.

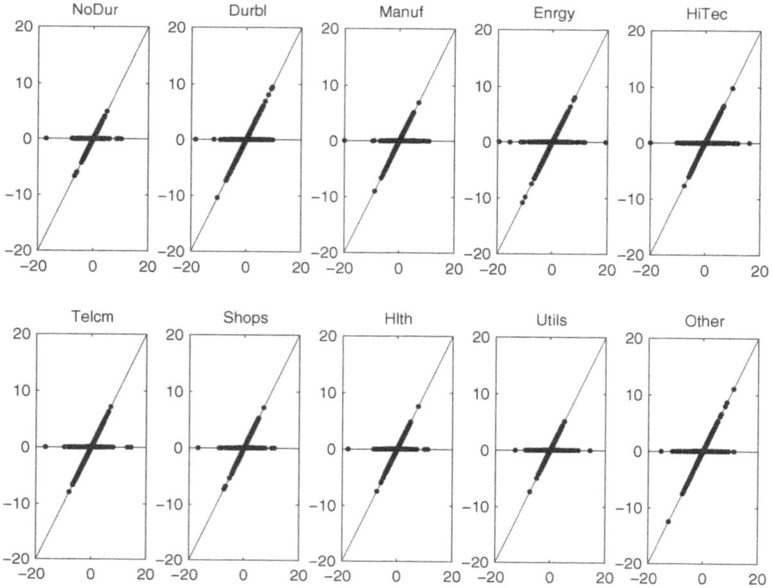

Figure 2.7 Scatter plot of BH versus MA returns: Industry portfolios.

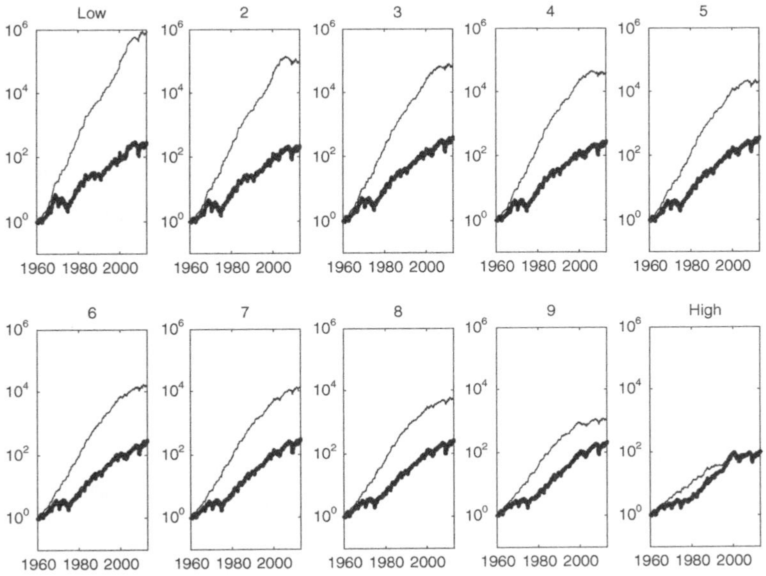

Figure 2.8 Cumulative returns of BH versus MA strategy: Size decile portfolios.

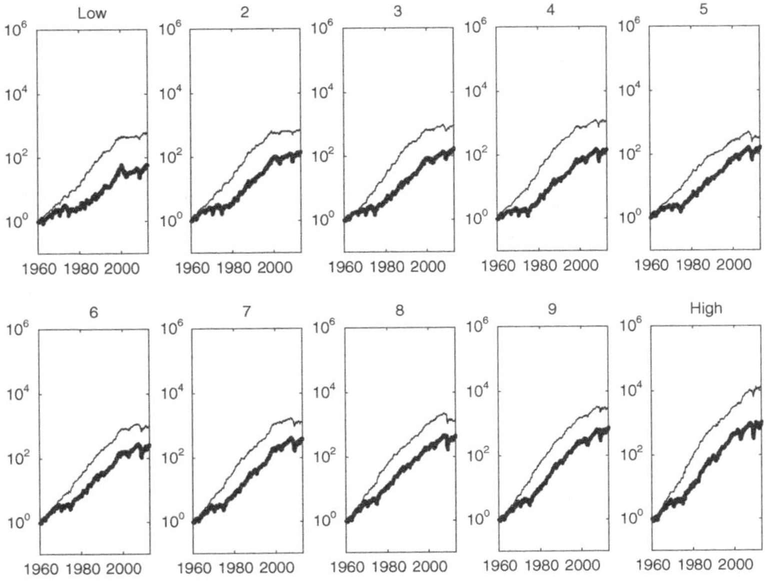

Figure 2.9 Cumulative returns of BH versus MA strategy: Book-to-market decile portfolios.

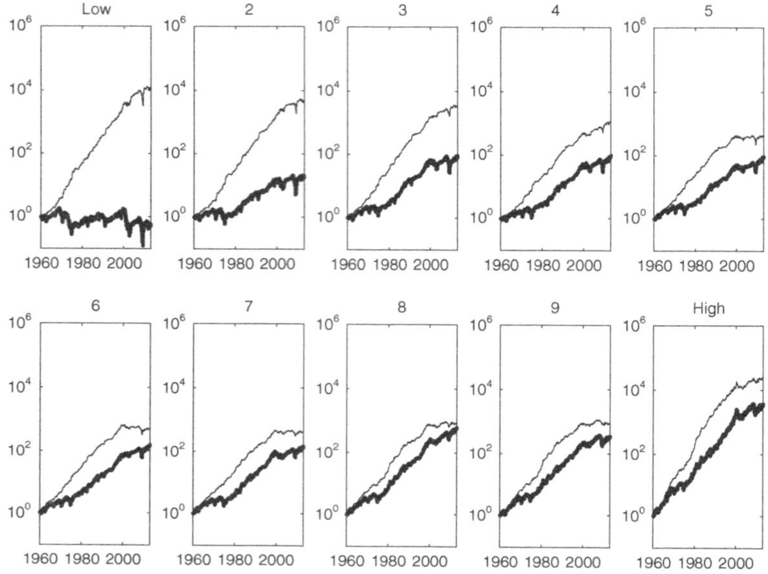

Figure 2.10 Cumulative returns of BH versus MA strategy: Momentum decile portfolios.

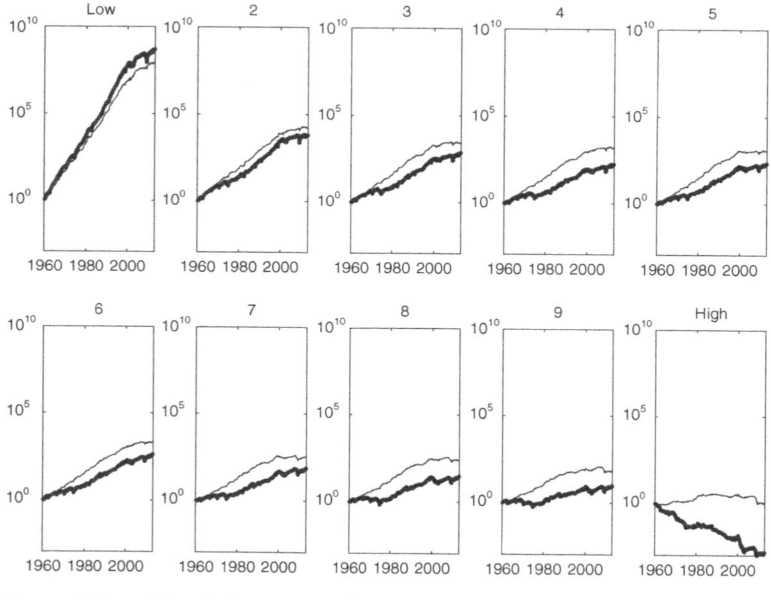

Figure 2.11 Cumulative returns of BH versus MA strategy: Short-term reversal decile portfolios.

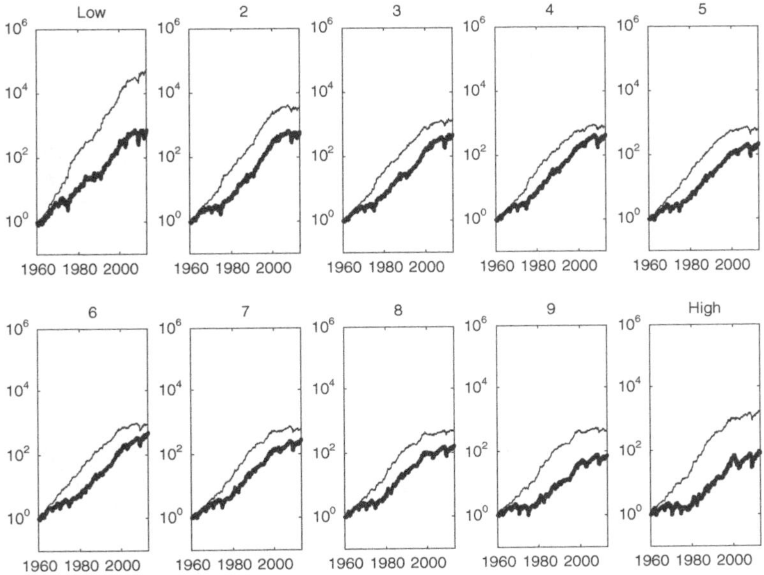

Figure 2.12 Cumulative returns of BH versus MA strategy: Long-term reversal decile portfolios.

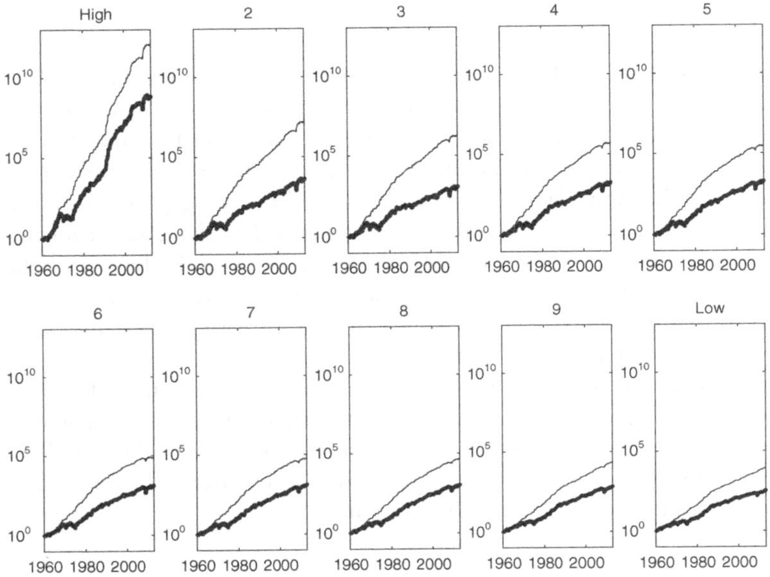

Figure 2.13 Cumulative returns of BH versus MA strategy: Volatility decile portfolios.

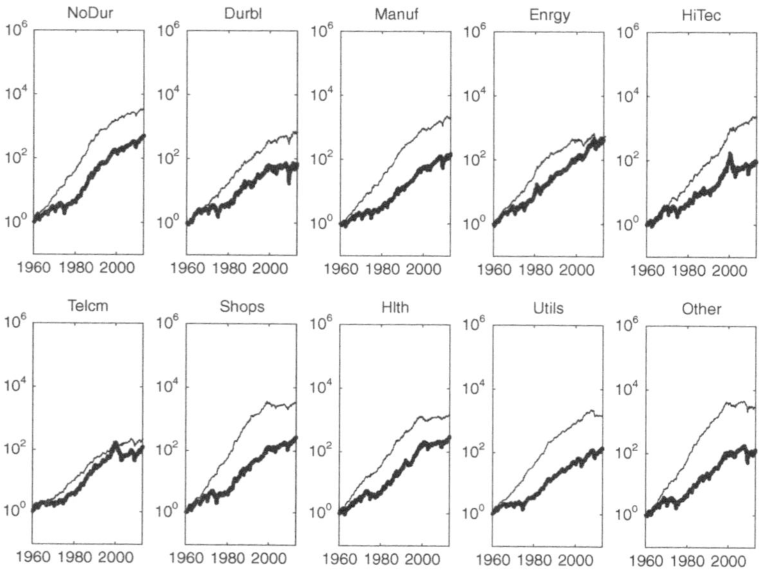

Figure 2.14 Cumulative returns of BH versus MA strategy: Industry portfolios.

few instances of a false-negative signal where we switch into the risk-free asset while the underlying risky asset has a positive excess return in the following period. Nevertheless, the signal is right about two out of every three times and in those instances the scatter plot resembles the payoff of an at-the-money put option combined with a long position in the underlying risky asset. This positive convexity is the driving factor for the relative outperformance of the MA strategy relative to the buy-and-hold strategy. Holding the signal success rate constant, risky assets with more volatile returns will experience a higher average outperformance and this is evidenced in all of the previous tables.

In order to develop this idea further, I introduce some additional notation. Suppose that the price of the underlying risky asset is S_0 which follows a geometric Brownian motion process. Black and Scholes (1973) then show that a European put option with a strike price of $X = S_0$ and T periods to maturity has the following value:

$$P_0 = e^{-rT} S_0 N(-d_2) - S_0 N(-d_1), \tag{2.9}$$

where

$$d_{1,2} = \frac{(r \pm \frac{\sigma^2}{2})T}{\sigma\sqrt{T}},$$

and $N(d)$ is the cumulative distribution of a standard Gaussian random variable. Now consider the value V_0 of an at-the-money protective put option combined with a long position in the underlying S_0:

$$V_0 = S_0\left[N(d_1) + e^{-rT}(1 - N(d_2))\right]. \tag{2.10}$$

Considering the replicating portfolio involved in this protective put strategy as if we had \$1 to invest at time 0 we need to invest a fraction w_S in the risky underlying security and a fraction w_B in the risk-free security:

$$w_S = \frac{N(d_1)}{N(d_1) + e^{-rT}(1 - N(d_2))},$$
$$w_B = \frac{e^{-rT}(1 - N(d_2))}{N(d_1) + e^{-rT}(1 - N(d_2))}.$$

Note that this strategy is always partially invested in the risky underlying and partially lends money at the risk-free rate of interest. If we believed completely the assumption regarding the data-generating process for the stock price then this would be a superior strategy to follow compared to the MA strategy. However, in light of the evidence against the log-normality of stock returns and volatility clustering, the MA strategy appears to provide a risky heuristic alternative where the switch is complete. We are always either fully 100% invested in the risky underlying or fully 100% invested in the risk-free security. In the absence of a strong belief regarding the data-generating process this is perhaps a suitable strategy to follow. Furthermore, if one is prepared to supply a more suitable process for the stock price and price the protective put, then the replicating portfolio in this case will provide a better alternative to the MA strategy.

Chapter 3

Performance Drivers

In this chapter, I investigate whether the superior returns of the MAP portfolios are due to their ability to time the market. Furthermore, I control the MAP performance for economic expansions and contractions as well as other state contingencies like the sign of the lagged market return. Finally, I investigate the conditional performance of the MAP returns while controlling for two instrumental variables with documented predictive power over stock returns and an additional risk factor to control the possible presence of liquidity risks.

3.1 Market Timing

The first approach toward testing for market-timing ability is the quadratic regression of Treynor and Mazuy (1966):

$$\mathrm{MAP}_{jt,L} = \alpha_j + \beta_{j,m} r_{mkt,t} + \beta_{j,m^2} r^2_{mkt,t} + \epsilon_{jt}, \quad j = 1, \ldots, N, \tag{3.1}$$

where statistically significant evidence of a positive β_{j,m^2} can be interpreted as evidence in favor of market-timing ability. The second approach is to allow for a state-contingent $\beta_{j,m}$ based on the direction of move of the market return as in Henriksson and Merton (1981):

$$\mathrm{MAP}_{jt,L} = \alpha_j + \beta_{j,m} r_{mkt,t} + \gamma_{j,m} r_{mkt,t} I_{\{r_{mkt,t}<0\}} + \epsilon_{jt}, \quad j = 1, \ldots, N, \tag{3.2}$$

where $I_{\{r_{mkt,t}<0\}}$ is an indicator function of the event of a negative market return. A statistically significant value of $\gamma_{j,m}$ is usually interpreted as evidence of successful market-timing ability.

Table 3.1 presents the results of the two market-timing regressions for various sets of value-weighted decile portfolios. Panel TM presents the empirical results from the Treynor and Mazuy (1966) quadratic regression while Panel HM presents the results for the state-contingent beta regression of Henriksson and Merton (1981). In both regressions, both β_{j,m^2} and $\gamma_{j,m}$ are highly statistically significant, indicating there is strong evidence of market-timing ability of the switching MA strategy. Nevertheless, the alphas of quite a few decile portfolios are also statistically significant at conventional levels. This suggests that market timing alone is not the sole driver of the abnormal returns generated by the switching MA strategy.

3.2 Business Cycles and Market States

Following Han, Yang and Zhou (2013), I investigate the performance of the MAP portfolio returns in economic expansions and contractions as well as in up and down markets as defined by the sign of the market return. Table 3.2 presents the results for the various sets of portfolio deciles. The evidence overwhelmingly indicates that MAP abnormal returns are higher during economic contractions and following positive market factor returns. For portfolios constructed by sorting on past performance (short-/long-term reversal and medium-term momentum) there is also evidence of a significant cross-sectional difference between the High and Low MAP abnormal returns which cannot be accounted for by the four Carhart (1997) factors and the recession dummy and up market dummy variables. This effect is smaller in magnitude than the one found by Han, Yang and Zhou (2013). Note, however, that they use daily equal-weighted returns which could potentially explain the difference in the cross-sectional results between this study and their study.

3.3 Conditional Models with Macroeconomic Variables

Ferson and Schadt (1996) make a strong case for using predetermined variables in controlling for changes in economic conditions while evaluating

Table 3.1 Market-timing regressions

This table reports alphas, betas, and adjusted R^2 of the market-timing regressions of the MAP excess returns on the market factor using portfolios sorted by various characteristics. The TM panel reports the results using the Treynor and Mazuy (1966) quadratic regression with the squared market factor (β_{m^2}) while the HM panel reports the results using the Henriksson and Merton (1981) regression with option-like returns on the market (γ_m). The sample period covers January 4, 1960 until December 31, 2013 with value-weighted portfolio returns. The length of the MA window is 20 days. Newey and West (1987) standard errors with five lags are used in reporting statistical significance of a two-sided null hypothesis at the 1%, 5%, and 10% level given by a ***, a**, and a*, respectively.

Portfolio	TM				HM			
	α	β_m	β_{m^2}	$\bar{R}^2$	α	β_m	γ_m	$\bar{R}^2$
Panel A: Size-sorted portfolios								
Low	13.284***	−0.429***	0.013***	0.411	3.962**	−0.356***	0.152***	0.411
2	12.286***	−0.559***	0.005***	0.454	6.788***	−0.519***	0.081***	0.454
3	10.144***	−0.571***	0.008***	0.484	3.928**	−0.524***	0.099***	0.484
4	10.342***	−0.570***	0.004***	0.495	6.829***	−0.544***	0.055***	0.495
5	9.420***	−0.575***	0.002	0.511	7.414***	−0.561***	0.030**	0.511
6	8.631***	−0.552***	0.004***	0.529	6.023***	−0.531***	0.044***	0.529
7	8.304***	−0.564***	0.004***	0.540	6.142***	−0.545***	0.039***	0.540
8	6.805***	−0.581***	0.005***	0.560	4.272***	−0.560***	0.045***	0.560
9	3.315***	−0.586***	0.010***	0.586	1.839	−0.566***	0.047***	0.584
High	0.533	−0.609***	0.008***	0.591	0.408	−0.599***	0.025**	0.589

continued

Table 3.1 Continued

Portfolio	TM				HM			
	α	β_m	β_{m^2}	$\bar{R}^2$	α	β_m	γ_m	$\bar{R}^2$
				Panel B: Book-to-market sorted portfolios				
Low	6.260***	−0.662***	0.003**	0.564	6.520***	−0.660***	0.005	0.564
2	4.239***	−0.597***	0.005***	0.566	4.075***	−0.591***	0.017	0.565
3	3.538***	−0.558***	0.009***	0.548	1.856	−0.538***	0.045***	0.547
4	3.024***	−0.568***	0.012***	0.551	0.676	−0.540***	0.065***	0.547
5	2.509**	−0.552***	0.005***	0.520	3.597**	−0.553***	0.000	0.520
6	3.121***	−0.532***	0.006***	0.519	2.623*	−0.522***	0.025**	0.518
7	2.316**	−0.509***	0.009***	0.503	1.370	−0.494***	0.037***	0.501
8	1.982	−0.527***	0.010***	0.472	1.176	−0.512***	0.038***	0.469
9	2.595**	−0.536***	0.009***	0.485	1.553	−0.520***	0.040***	0.483
High	4.901***	−0.578***	0.009***	0.460	3.102*	−0.557***	0.048***	0.458
				Panel C: Momentum-sorted portfolios				
Low	22.658***	−0.834***	−0.006***	0.451	23.738***	−0.847***	−0.031	0.451
2	13.787***	−0.700***	−0.004***	0.494	14.144***	−0.706***	−0.017	0.494
3	9.502***	−0.605***	−0.001	0.511	8.638***	−0.601***	0.007	0.511
4	5.536***	−0.588***	0.007***	0.535	3.588**	−0.569***	0.044***	0.534
5	5.201***	−0.560***	−0.000	0.528	6.685***	−0.569***	−0.018	0.528
6	3.243***	−0.538***	0.005***	0.522	4.942***	−0.543***	−0.005	0.521
7	2.349**	−0.543***	0.008***	0.534	1.962	−0.532***	0.028**	0.532
8	0.055	−0.536***	0.012***	0.517	−1.645	−0.513***	0.056***	0.513
9	1.684	−0.591***	0.010***	0.528	−1.017	−0.563***	0.063***	0.526
High	3.805**	−0.704***	0.010***	0.475	1.651	−0.680***	0.055***	0.474

Panel D: Short-term reversal sorted portfolios								
Low	2.136	−0.838***	−0.012***	0.393	11.532***	−0.910***	−0.149***	0.393
2	5.424***	−0.758***	−0.003**	0.495	9.119***	−0.785***	−0.054***	0.495
3	5.656***	−0.670***	−0.002*	0.510	8.747***	−0.692***	−0.044***	0.510
4	5.159***	−0.621***	0.006***	0.535	6.319***	−0.622***	0.004	0.534
5	3.852***	−0.593***	0.008***	0.553	3.086**	−0.580***	0.033***	0.552
6	3.090***	−0.557***	0.010***	0.542	2.098	−0.541***	0.040***	0.539
7	3.426***	−0.554***	0.008***	0.559	2.463*	−0.540***	0.034***	0.558
8	4.351***	−0.537***	0.008***	0.548	0.616	−0.505***	0.069***	0.547
9	4.035***	−0.544***	0.007***	0.530	1.062	−0.517***	0.057***	0.530
High	13.045***	−0.658***	0.007***	0.525	4.552**	−0.597***	0.123***	0.527
Panel E: Long-term reversal sorted portfolios								
Low	7.855***	−0.674***	0.012***	0.486	4.265**	−0.639***	0.078***	0.484
2	3.321**	−0.592***	0.010***	0.512	3.218*	−0.580***	0.032**	0.510
3	1.738	−0.539***	0.010***	0.507	0.543	−0.521***	0.045***	0.504
4	0.222	−0.531***	0.013***	0.508	−0.097	−0.516***	0.041***	0.504
5	0.973	−0.527***	0.014***	0.513	0.035	−0.507***	0.052***	0.508
6	2.143**	−0.506***	0.005***	0.511	2.299	−0.502***	0.013	0.510
7	2.172**	−0.518***	0.006***	0.517	0.625	−0.502***	0.037***	0.516
8	2.553**	−0.565***	0.008***	0.542	0.098	−0.541***	0.055***	0.541
9	5.624***	−0.620***	0.002	0.538	4.384***	−0.611***	0.020	0.538
High	9.471***	−0.767***	−0.004***	0.543	9.705***	−0.773***	−0.014	0.543

continued

Table 3.1 Continued

Portfolio	TM				HM			
	α	β_m	β_{m^2}	$\bar{R}^2$	α	β_m	γ_m	$\bar{R}^2$
				Panel F: Volatility-sorted portfolios				
High	13.566***	−0.547***	0.011***	0.355	6.236***	−0.490***	0.120***	0.355
2	14.701***	−0.644***	0.012***	0.464	6.367***	−0.579***	0.135***	0.465
3	14.487***	−0.628***	0.006***	0.481	10.014***	−0.593***	0.071***	0.481
4	11.485***	−0.582***	0.005***	0.486	8.441***	−0.557***	0.052***	0.486
5	10.269***	−0.533***	0.005***	0.485	6.411***	−0.503***	0.062***	0.486
6	8.249***	−0.496***	0.006***	0.491	5.574***	−0.473***	0.049***	0.490
7	7.401***	−0.440***	0.007***	0.486	4.460***	−0.414***	0.055***	0.485
8	6.144***	−0.379***	0.011***	0.475	1.787	−0.340***	0.085***	0.472
9	5.574***	−0.299***	0.011***	0.423	0.573	−0.256***	0.092***	0.419
Low	4.747***	−0.169***	0.010***	0.257	0.768	−0.133***	0.077***	0.250
				Panel G: Industry-sorted portfolios				
NoDur	2.900***	−0.429***	0.011***	0.440	−0.345	−0.397***	0.070***	0.437
Durbl	5.008***	−0.673***	0.007***	0.466	5.146**	−0.666***	0.018	0.465
Manuf	4.916***	−0.587***	0.010***	0.546	3.398**	−0.567***	0.048***	0.544
Enrgy	1.433	−0.552***	0.004***	0.338	2.840	−0.557***	−0.005	0.338
HiTec	8.243***	−0.784***	0.003*	0.491	9.656***	−0.790***	−0.009	0.491
Telcm	2.958**	−0.539***	0.001	0.413	5.396***	−0.552***	−0.025*	0.414
Shops	5.698***	−0.548***	0.005***	0.468	5.603***	−0.542***	0.017	0.468
Hlth	3.029**	−0.521***	0.009***	0.416	2.270	−0.506***	0.035**	0.415
Utils	5.661***	−0.348***	0.001	0.300	4.986***	−0.343***	0.012	0.300
Other	7.099***	−0.647***	0.005***	0.537	4.680***	−0.626***	0.045***	0.537

Table 3.2 Factor regressions with business cycles and down markets

This table reports alphas, betas, and adjusted R^2 of the factor regressions of the MAP excess returns using the Carhart four-factor model with NBER recession indicator dummy variable (RI) and down market indicators (DN) using portfolios sorted by various characteristics. Alphas are annualized and in percent. The sample period covers January 4, 1960 until December 31, 2013 with value-weighted portfolio returns. The length of the MA window is 20 days. Newey and West (1987) standard errors with 24 lags are used in reporting statistical significance of a two-sided null hypothesis at the 1%, 5%, and 10% level given by a***, a**, and a*, respectively.

Portfolio	Recession Dummy							Down Market Dummy						
	α	β_m	β_s	β_h	β_u	*RI*	$\bar{R}^2$	α	β_m	β_s	β_h	β_u	*DN*	$\bar{R}^2$
Panel A: Size sorted portfolios														
Low	17.044***	−0.503***	−0.523***	−0.164***	−0.086***	0.053***	0.562	43.284***	−0.573***	−0.522***	−0.155***	−0.095***	−0.205***	0.574
2	15.053***	−0.638***	−0.601***	−0.181***	−0.088***	0.033**	0.591	45.899***	−0.723***	−0.599***	−0.171***	−0.099***	−0.250***	0.604
3	12.840***	−0.644***	−0.557***	−0.156***	−0.081***	0.046***	0.602	45.303***	−0.733***	−0.555***	−0.146***	−0.093***	−0.260***	0.616
4	12.277***	−0.633***	−0.502***	−0.118***	−0.093***	0.038***	0.596	44.435***	−0.721***	−0.500***	−0.107***	−0.104***	−0.260***	0.610
5	10.330***	−0.626***	−0.451***	−0.077***	−0.075***	0.036***	0.593	43.789***	−0.718***	−0.449***	−0.066***	−0.087***	−0.271***	0.609
6	9.498***	−0.590***	−0.317***	−0.049***	−0.079***	0.043***	0.578	42.712***	−0.681***	−0.314***	−0.039***	−0.091***	−0.267***	0.594
7	9.414***	−0.602***	−0.258***	−0.071***	−0.085***	0.040***	0.574	43.673***	−0.696***	−0.256***	−0.059***	−0.097***	−0.277***	0.592
8	7.988***	−0.615***	−0.181***	−0.081***	−0.080***	0.039***	0.579	44.263***	−0.715***	−0.179***	−0.069***	−0.093***	−0.294***	0.598
9	5.345***	−0.615***	−0.063***	−0.093***	−0.079***	0.043***	0.591	43.507***	−0.720***	−0.060***	−0.080***	−0.092***	−0.309***	0.613
High	1.845*	−0.600***	0.145***	0.043***	−0.047***	0.014	0.600	38.014***	−0.702***	0.147***	0.056***	−0.060***	−0.302***	0.620

continued

Table 3.2 Continued

Portfolio	Recession Dummy							Down Market Dummy						
	α	β_m	β_s	β_h	β_u	RI	$\bar{R}^2$	α	β_m	β_s	β_h	β_u	DN	$\bar{R}^2$
Panel B: Book-to-market sorted portfolios														
Low	4.157***	−0.620***	0.038***	0.306***	−0.041***	0.037***	0.593	37.801***	−0.713***	0.041***	0.317***	−0.053***	−0.273***	0.606
2	4.508***	−0.593***	0.023***	0.079***	−0.071***	0.030**	0.572	38.306***	−0.687***	0.025***	0.090***	−0.083***	−0.276***	0.589
3	5.813***	−0.569***	0.023***	0.000	−0.099***	0.016	0.554	38.530***	−0.661***	0.026***	0.012	−0.111***	−0.272***	0.572
4	6.237***	−0.602***	−0.007	−0.153***	−0.068***	0.031***	0.556	41.824***	−0.701***	−0.004	−0.141***	−0.081***	−0.291***	0.575
5	5.032***	−0.587***	−0.011	−0.198***	−0.056***	0.004	0.533	38.106***	−0.682***	−0.009	−0.187***	−0.068***	−0.279***	0.552
6	5.710***	−0.573***	−0.072***	−0.198***	−0.055***	0.017	0.534	37.365***	−0.662***	−0.070***	−0.187***	−0.067***	−0.262***	0.551
7	6.001***	−0.562***	−0.032***	−0.280***	−0.081***	0.019	0.531	35.630***	−0.645***	−0.030***	−0.270***	−0.091***	−0.245***	0.547
8	6.568***	−0.602***	−0.075***	−0.442***	−0.033***	0.014	0.536	38.481***	−0.692***	−0.072***	−0.432***	−0.045***	−0.266***	0.552
9	7.604***	−0.611***	−0.064***	−0.409***	−0.095***	0.010	0.539	38.159***	−0.698***	−0.062***	−0.399***	−0.106***	−0.255***	0.554
High	10.036***	−0.669***	−0.185***	−0.462***	−0.090***	0.017	0.523	39.030***	−0.750***	−0.183***	−0.452***	−0.101***	−0.240***	0.534
Panel C: Momentum-sorted portfolios														
Low	16.241***	−0.818***	−0.289***	−0.042**	0.435***	0.047**	0.523	51.933***	−0.916***	−0.287***	−0.030*	0.423***	−0.287***	0.530
2	8.160***	−0.680***	−0.093***	−0.026**	0.309***	0.056***	0.543	43.649***	−0.776***	−0.091***	−0.014	0.297***	−0.282***	0.554
3	5.451***	−0.597***	−0.010	−0.063***	0.190***	0.063***	0.539	36.180***	−0.679***	−0.008	−0.053***	0.179***	−0.239***	0.549
4	5.368***	−0.597***	0.015*	−0.096***	0.072***	0.043***	0.542	40.008***	−0.692***	0.017*	−0.084***	0.059***	−0.279***	0.559
5	4.857***	−0.570***	−0.003	−0.072***	0.002	0.018	0.530	37.827***	−0.662***	−0.000	−0.061***	−0.010	−0.273***	0.547
6	4.441***	−0.556***	0.024***	−0.083***	−0.061***	0.027**	0.526	36.510***	−0.646***	0.027***	−0.072***	−0.073***	−0.263***	0.543
7	5.376***	−0.572***	0.012	−0.103***	−0.131***	0.015	0.547	38.187***	−0.665***	0.014*	−0.092***	−0.143***	−0.273***	0.566
8	4.767***	−0.578***	0.021**	−0.139***	−0.204***	0.018	0.547	37.370***	−0.670***	0.023***	−0.128***	−0.216***	−0.270***	0.565
9	6.362***	−0.640***	−0.074***	−0.105***	−0.261***	0.021*	0.573	40.917***	−0.737***	−0.071***	−0.093***	−0.274***	−0.286***	0.590
High	9.337***	−0.757***	−0.226***	0.028**	−0.409***	0.020	0.566	50.742***	−0.874***	−0.223***	0.042***	−0.424***	−0.344***	0.581

Panel D: Short-term reversal sorted portfolios													
Low	−2.145	−0.840***	−0.239***	0.012	0.084***	0.025	0.401	66.028***	−1.033***	−0.234***	0.035*	0.059***	−0.569***	0.427
2	3.834**	−0.768***	−0.098***	−0.029**	−0.002	0.030*	0.497	57.612***	−0.919***	−0.094***	−0.011	−0.022**	−0.445***	0.521
3	5.627***	−0.687***	−0.051***	−0.060***	−0.062***	0.011	0.513	49.210***	−0.811***	−0.048***	−0.045***	−0.078***	−0.365***	0.534
4	6.504***	−0.640***	−0.009	−0.062***	−0.079***	0.030**	0.538	47.023***	−0.753***	−0.006	−0.049***	−0.094***	−0.333***	0.559
5	6.138***	−0.614***	−0.012	−0.071***	−0.070***	0.018	0.555	45.638***	−0.725***	−0.009	−0.058***	−0.084***	−0.329***	0.579
6	5.996***	−0.579***	−0.006	−0.077***	−0.076***	0.013	0.544	42.718***	−0.683***	−0.003	−0.065***	−0.089***	−0.307***	0.567
7	5.280***	−0.574***	−0.013	−0.075***	−0.063***	0.025**	0.562	38.147***	−0.666***	−0.011	−0.064***	−0.075***	−0.270***	0.580
8	6.950***	−0.559***	−0.018**	−0.066***	−0.092***	0.015	0.554	33.919***	−0.635***	−0.016*	−0.056***	−0.102***	−0.223***	0.567
9	6.396***	−0.566***	−0.047***	−0.080***	−0.046***	0.007	0.533	28.448***	−0.628***	−0.046***	−0.073***	−0.054***	−0.185***	0.541
High	14.795***	−0.678***	−0.186***	−0.027**	−0.002	0.011	0.535	26.409***	−0.711***	−0.185***	−0.023*	−0.006	−0.095***	0.536
Panel E: Long-term reversal sorted portfolios														
Low	13.834***	−0.752***	−0.336***	−0.260***	−0.139***	0.002	0.529	48.216***	−0.851***	−0.333***	−0.248***	−0.152***	−0.290***	0.542
2	8.100***	−0.651***	−0.149***	−0.224***	−0.131***	0.009	0.537	40.826***	−0.744***	−0.147***	−0.213***	−0.143***	−0.274***	0.552
3	5.749***	−0.589***	−0.060***	−0.212***	−0.114***	0.019	0.525	36.994***	−0.676***	−0.057***	−0.201***	−0.125***	−0.258***	0.541
4	3.752***	−0.577***	−0.029***	−0.225***	−0.072***	0.036***	0.522	34.609***	−0.662***	−0.027***	−0.215***	−0.083***	−0.249***	0.537
5	4.934***	−0.572***	−0.006	−0.225***	−0.073***	0.032***	0.527	36.527***	−0.660***	−0.003	−0.215***	−0.084***	−0.257***	0.543
6	4.274***	−0.532***	0.047***	−0.135***	−0.086***	0.013	0.523	33.773***	−0.615***	0.050***	−0.125***	−0.096***	−0.245***	0.539
7	3.821***	−0.537***	0.044***	−0.085***	−0.080***	0.025**	0.524	33.447***	−0.620***	0.046***	−0.075***	−0.090***	−0.243***	0.539
8	4.562***	−0.581***	0.057***	−0.048***	−0.100***	0.028**	0.549	37.750***	−0.673***	0.059***	−0.037***	−0.112***	−0.272***	0.566
9	4.980***	−0.624***	0.054***	0.004	−0.083***	0.043***	0.543	39.393***	−0.718***	0.056***	0.016	−0.096***	−0.277***	0.558
High	6.079***	−0.735***	−0.069***	0.258***	−0.023***	0.035**	0.560	47.647***	−0.851***	−0.066***	0.272***	−0.038***	−0.340***	0.574

continued

Table 3.2 Continued

Portfolio	Recession Dummy							Down Market Dummy						
	α	β_m	β_s	β_h	β_u	RI	$\bar{R}^2$	α	β_m	β_s	β_h	β_u	DN	$\bar{R}^2$
Panel F: Volatility-sorted portfolios														
High	17.221***	−0.620***	−0.533***	−0.169***	−0.061***	0.035**	0.439	51.260***	−0.714***	−0.531***	−0.158***	−0.073***	−0.277***	0.452
2	17.438***	−0.713***	−0.504***	−0.164***	−0.047***	0.059***	0.535	54.472***	−0.813***	−0.502***	−0.152***	−0.060***	−0.294***	0.548
3	16.075***	−0.695***	−0.453***	−0.192***	−0.065***	0.057***	0.549	51.445***	−0.791***	−0.451***	−0.180***	−0.078***	−0.281***	0.562
4	12.960***	−0.646***	−0.381***	−0.210***	−0.055***	0.052***	0.546	49.398***	−0.745***	−0.379***	−0.198***	−0.068***	−0.291***	0.563
5	11.708***	−0.594***	−0.308***	−0.222***	−0.055***	0.052***	0.539	45.717***	−0.686***	−0.306***	−0.211***	−0.068***	−0.271***	0.555
6	9.845***	−0.551***	−0.240***	−0.216***	−0.053***	0.048***	0.534	43.962***	−0.644***	−0.238***	−0.205***	−0.065***	−0.273***	0.554
7	9.166***	−0.485***	−0.165***	−0.174***	−0.062***	0.042***	0.516	39.795***	−0.568***	−0.164***	−0.164***	−0.073***	−0.245***	0.536
8	8.856***	−0.419***	−0.112***	−0.153***	−0.062***	0.039***	0.495	37.403***	−0.497***	−0.110***	−0.144***	−0.073***	−0.229***	0.518
9	7.904***	−0.333***	−0.073***	−0.137***	−0.050***	0.041***	0.438	31.466***	−0.396***	−0.071***	−0.130***	−0.058***	−0.186***	0.459
Low	6.515***	−0.188***	−0.022***	−0.062***	−0.047***	0.035***	0.258	19.057***	−0.220***	−0.022***	−0.058***	−0.052***	−0.095***	0.267
Panel G: Industry-sorted portfolios														
NoDur	3.026***	−0.408***	0.066***	−0.018*	0.014**	−0.002	0.427	35.932***	−0.503***	0.068***	−0.006	0.002	−0.280***	0.454
Durbl	4.693***	−0.716***	−0.066***	−0.231***	0.183***	0.045**	0.509	43.275***	−0.822***	−0.064***	−0.217***	0.170***	−0.312***	0.522
Manuf	2.438*	−0.592***	−0.013	−0.074***	0.092***	0.040***	0.549	44.700***	−0.709***	−0.011	−0.060***	0.077***	−0.345***	0.574
Enrgy	−0.170	−0.563***	0.116***	−0.187***	0.062***	0.016	0.366	45.463***	−0.692***	0.119***	−0.171***	0.045***	−0.381***	0.389
HiTec	−1.027	−0.674***	−0.019	0.542***	0.267***	0.032*	0.543	42.611***	−0.795***	−0.016	0.558***	0.251***	−0.359***	0.558
Telcm	1.591	−0.503***	0.063***	−0.066***	0.176***	0.006	0.411	38.459***	−0.608***	0.066***	−0.053***	0.163***	−0.310***	0.431
Shops	2.528*	−0.527***	−0.010	0.033***	0.076***	0.006	0.456	34.974***	−0.619***	−0.008	0.045***	0.064***	−0.273***	0.472
Hlth	1.867	−0.478***	0.059***	0.122***	0.030***	−0.011	0.409	33.305***	−0.569***	0.061***	0.133***	0.018**	−0.270***	0.425
Utils	3.423***	−0.375***	0.059***	−0.241***	0.033***	0.010	0.345	34.289***	−0.462***	0.061***	−0.230***	0.022***	−0.258***	0.368
Other	4.072***	−0.702***	−0.146***	−0.283***	0.232***	0.033**	0.604	52.218***	−0.837***	−0.143***	−0.267***	0.214***	−0.397***	0.630

investment performance. I augment the four-factor Carhart (1997) model with an intercept that is a linear function of a set of instruments as well as cross-products of the instrumental variables with the market return to allow for state-dependent betas with the market factor. I use investor sentiment due to Baker and Wurgler (2006), the aggregate liquidity factor of Pastor and Stambaugh (2003), and the default spread of Moody's BAA corporate bond yield over the AAA corporate bond yield as the instrumental variables Z_t in the following regression:

$$\begin{aligned}\mathrm{MAP}_{jt,L} &= \alpha_j + \beta_{j,m} r_{mkt,t} + \beta_{j,s} r_{smb,t} + \beta_{j,h} r_{hml,t} + \beta_{j,u} r_{umd,t} \\ &\quad + \beta_{j,Z} Z_{t-1} + \gamma_{j,Z} Z_{t-1} r_{mkt,t} + \epsilon_{jt}. \end{aligned} \tag{3.3}$$

Table 3.3 presents the results of the conditional model estimation. Changes in investor sentiment are important in increasing conditional alphas but also lead to higher betas with the market factor as evidenced by the positive coefficient estimate of the cross-product variable $\Delta S \times r_m$. Increases in the default spread result in higher conditional alphas but lower conditional betas with the market. The evidence for the aggregate liquidity factor is a little mixed and there appear to be some cross-sectional differences between the various decile portfolio returns. However, all the unconditional alphas for all sets of portfolios are highly statistically and economically significant. This suggests that investor sentiment, liquidity and especially the default premium can account for the MAP abnormal returns, at least using this particular conditional specification.

Finally, I put all the instrumental variables along with an National Bureau of Economic Research (NBER) recession dummy variable in the same regression with the four Carhart (1997) factors as well as interactions between the instrumental variables and the market return. Table 3.4 presents the results from this conditional model specification. The previous results vis-à-vis investor sentiment, the default spread, and liquidity largely hold with the same signs albeit with a smaller degree of statistical significance. The recession indicator emerges as an important driver of conditional market betas where for all sets of portfolios the interaction term $RI \times r_m$ is always negative and highly statistically significant. This suggests that for almost all portfolios betas with the market tend to be significantly lower during economic recessions compared to their values during economic expansions.

Table 3.3 Conditional regressions with market dividend yield and treasury bill rate

This table reports alphas, betas, and adjusted R^2 of the market timing regressions of the MAP excess returns on the CAPM market factor plus one instrumental variable and an interaction term of the instrumental variable with the market's excess return using portfolios sorted by various characteristics. Alphas are annualized and in percent. The sample period covers January 4, 1960 until December 31, 2013. The length of the moving-average window is 20 days. Newey and West (1987) standard errors with five lags are used in reporting statistical significance of a two-sided null hypothesis at the 1%, 5%, and 10% level given by a***, a**, and a*, respectively.

Portfolio	**Market Dividend Yield**					**Treasury Bill Rate**				
	α	DP_m	β_m	$DP_m \times r_m$	$\bar{R}^2$	α	r_f	β_m	$r_f \times r_m$	$\bar{R}^2$
				Panel A: Size-sorted portfolios						
Low	15.574***	0.001	−0.538***	0.039***	0.409	8.015***	1.831***	−0.497***	3.690***	0.412
2	8.541**	0.007	−0.768***	0.078***	0.464	4.133	2.032***	−0.678***	6.991***	0.466
3	5.581	0.008*	−0.805***	0.087***	0.496	4.119	1.717***	−0.690***	6.894***	0.495
4	1.906	0.012**	−0.786***	0.081***	0.506	1.485	2.074***	−0.661***	5.297***	0.502
5	0.818	0.012**	−0.826***	0.095***	0.527	1.980	1.650***	−0.657***	4.876***	0.517
6	1.100	0.011**	−0.757***	0.077***	0.541	1.658	1.640***	−0.597***	2.594***	0.531
7	0.158	0.012***	−0.721***	0.059***	0.547	2.775	1.366***	−0.609***	2.550***	0.543
8	−3.291	0.015***	−0.751***	0.063***	0.568	1.506	1.355***	−0.627***	2.615***	0.562
9	−7.999**	0.018***	−0.738***	0.055***	0.591	0.844	0.994***	−0.632***	2.429***	0.586
High	−1.728	0.005	−0.735***	0.046***	0.594	3.131	−0.182	−0.600***	−0.790**	0.590

Panel B: Book-to-market sorted portfolios										
Low	−5.984	0.017***	−0.752***	0.034***	0.567	1.312	1.082***	−0.589***	−4.432***	0.569
2	−5.321	0.014***	−0.683***	0.031***	0.568	2.404	0.593	−0.582***	−1.158***	0.566
3	−5.752*	0.015***	−0.622***	0.023***	0.548	−0.044	1.133***	−0.543***	−1.116***	0.547
4	−6.545*	0.016***	−0.651***	0.029***	0.549	1.836	0.835**	−0.618***	2.588***	0.549
5	−5.025	0.011**	−0.656***	0.038***	0.524	1.012	0.547	−0.619***	3.853***	0.525
6	−2.804	0.010**	−0.669***	0.051***	0.524	1.395	0.686*	−0.621***	5.224***	0.527
7	−4.767	0.012***	−0.596***	0.031***	0.503	0.415	0.843**	−0.595***	4.939***	0.509
8	−2.669	0.009*	−0.611***	0.030***	0.471	1.302	0.645	−0.671***	8.503***	0.489
9	−0.484	0.007	−0.651***	0.042***	0.486	3.485	0.322	−0.663***	7.407***	0.498
High	3.332	0.005	−0.658***	0.029***	0.459	5.907**	0.311	−0.719***	8.371***	0.474
Panel C: Momentum-sorted portfolios										
Low	23.865***	−0.004	−1.156***	0.123***	0.462	15.027***	1.375*	−0.964***	8.032***	0.457
2	9.397**	0.004	−0.948***	0.095***	0.504	7.831***	1.072**	−0.827***	7.771***	0.504
3	2.852	0.008	−0.758***	0.058***	0.516	5.595**	0.789*	−0.694***	5.357***	0.518
4	0.365	0.009**	−0.693***	0.038***	0.537	6.021***	0.281	−0.656***	3.908***	0.538
5	−6.316*	0.015***	−0.652***	0.035***	0.531	−0.383	1.166***	−0.610***	3.014***	0.531
6	−10.939***	0.020***	−0.582***	0.015***	0.523	−1.719	1.252***	−0.566***	1.472***	0.523
7	−6.159*	0.014***	−0.582***	0.013***	0.534	−1.054	1.074***	−0.558***	0.634*	0.534
8	−10.673***	0.018***	−0.585***	0.016***	0.514	−2.304	1.033***	−0.544***	0.130	0.513
9	−8.727**	0.017***	−0.657***	0.022***	0.528	−1.932	1.210***	−0.587***	−0.625	0.527
High	−10.027**	0.021***	−0.858***	0.057***	0.478	0.700	1.081**	−0.662***	−2.838***	0.476

continued

Table 3.3 Continued

Portfolio	Market Dividend Yield					Treasury Bill Rate				
	α	DP_m	β_m	$DP_m \times r_m$	$\bar{R}^2$	α	r_f	β_m	$r_f \times r_m$	$\bar{R}^2$
Panel D: Short-term reversal sorted portfolios										
Low	−1.874	0.000	−1.305***	0.178***	0.414	0.219	−0.148	−0.944***	6.511***	0.397
2	−4.417	0.011*	−1.034***	0.105***	0.507	3.324	0.265	−0.821***	3.862***	0.498
3	−7.377*	0.016***	−0.796***	0.048***	0.514	−0.412	1.108**	−0.701***	1.929***	0.511
4	−4.365	0.014***	−0.755***	0.050***	0.538	1.023	1.125***	−0.640***	0.989**	0.534
5	−3.669	0.012***	−0.732***	0.051***	0.557	1.769	0.815**	−0.642***	2.709***	0.554
6	−2.055	0.010**	−0.659***	0.037***	0.543	0.924	0.922**	−0.599***	2.165***	0.542
7	−7.637**	0.017***	−0.648***	0.034***	0.561	−0.169	1.101***	−0.587***	1.721***	0.559
8	−1.592	0.010**	−0.632***	0.034***	0.551	1.158	1.051***	−0.564***	1.318***	0.549
9	−4.057	0.013***	−0.609***	0.023***	0.532	0.080	1.194***	−0.581***	1.991***	0.532
High	13.240***	0.001	−0.785***	0.048***	0.528	8.051***	1.336***	−0.709***	3.034***	0.527
Panel E: Long-term reversal sorted portfolios										
Low	7.076	0.004	−0.805***	0.048***	0.486	9.142***	0.308	−0.790***	6.740***	0.491
2	−1.979	0.010**	−0.679***	0.031***	0.512	4.348*	0.312	−0.664***	3.995***	0.514
3	−7.654**	0.016***	−0.582***	0.014***	0.506	1.072	0.675*	−0.594***	3.034***	0.507
4	−7.851**	0.015***	−0.581***	0.017***	0.505	−0.230	0.711*	−0.605***	4.111***	0.509
5	−7.719**	0.016***	−0.565***	0.011***	0.508	0.115	0.859**	−0.586***	3.113***	0.511
6	−9.980***	0.018***	−0.528***	0.007*	0.512	−0.883	0.844**	−0.532***	1.373***	0.512
7	−8.319**	0.016***	−0.545***	0.009**	0.518	−0.802	0.925**	−0.532***	0.622*	0.517
8	−5.002	0.013***	−0.626***	0.021***	0.542	2.357	0.464	−0.567***	−0.200	0.541
9	−10.205***	0.022***	−0.658***	0.014***	0.539	0.861	1.034**	−0.632***	0.597	0.539
High	−6.098	0.019***	−0.939***	0.066***	0.549	2.286	1.247**	−0.746***	−1.233***	0.544

Panel F: Volatility-sorted portfolios										
High	18.360***	−0.003	−0.568***	0.007	0.351	14.068***	0.475	−0.586***	2.159***	0.352
2	14.399***	0.004	−0.691***	0.016***	0.462	11.818***	1.245**	−0.695***	2.769***	0.463
3	10.404**	0.008	−0.654***	0.009	0.480	8.893***	1.490***	−0.681***	3.132***	0.482
4	5.145	0.010*	−0.641***	0.022***	0.485	6.070**	1.419***	−0.657***	4.456***	0.490
5	1.980	0.013**	−0.577***	0.016***	0.486	4.696*	1.443***	−0.610***	4.504***	0.491
6	−0.478	0.014***	−0.553***	0.021***	0.491	2.460	1.528***	−0.570***	4.296***	0.496
7	2.265	0.009**	−0.473***	0.011***	0.485	3.544*	1.157***	−0.484***	2.466***	0.487
8	2.918	0.008**	−0.377***	−0.003	0.469	5.396***	0.728**	−0.408***	1.440***	0.470
9	5.618**	0.004	−0.252***	−0.020***	0.416	6.887***	0.289	−0.301***	−0.173	0.414
Low	4.990**	0.003	−0.054***	−0.045***	0.262	5.821***	0.257	−0.125***	−2.872***	0.254
Panel G: Industry-sorted portfolios										
NoDur	−4.437	0.013***	−0.340***	−0.036***	0.440	1.897	0.710*	−0.405***	−1.723***	0.438
Durbl	3.175	0.005	−0.750***	0.028***	0.468	3.100	0.818	−0.759***	5.058***	0.471
Manuf	−6.647*	0.018***	−0.587***	−0.001	0.543	2.405	0.965**	−0.607***	1.023***	0.543
Enrgy	−16.161***	0.025***	−0.496***	−0.022***	0.339	−3.884	1.258**	−0.608***	3.194***	0.340
HiTec	8.698*	−0.000	−1.127***	0.130***	0.506	6.351**	0.503	−0.725***	−3.573***	0.493
Telcm	3.110	−0.001	−0.828***	0.110***	0.434	5.237**	−0.367	−0.653***	6.855***	0.426
Shops	−11.397***	0.024***	−0.574***	0.008	0.471	0.354	1.316***	−0.528***	−1.489***	0.471
Hlth	−8.201*	0.018***	−0.403***	−0.046***	0.419	4.851*	0.005	−0.469***	−3.293***	0.418
Utils	−3.623	0.013***	−0.388***	0.014***	0.302	1.737	0.889**	−0.431***	4.879***	0.311
Other	−8.111**	0.022***	−0.838***	0.071***	0.545	2.321	1.286***	−0.743***	5.557***	0.544

Table 3.4 Conditional regressions with market dividend yield, recession dummy and treasury bill rate

This table reports alphas, betas, and adjusted R^2 of the market timing regressions of the MAP excess returns on the CAPM market factor with three instrumental variables (the dividend yield on the market, DP_m, a recession dummy RI, and the treasury bill rate r_f) as well as interaction terms of the instrumental variable with the market's excess return using portfolios sorted by various characteristics. Alphas are annualized and in percent. The sample period covers January 4, 1960 until December 31, 2013. The length of the MA window is 20 days. Newey and West (1987) standard errors with five lags are used in reporting statistical significance of a two-sided null hypothesis at the 1%, 5%, and 10% level given by a***, a**, and a*, respectively.

Portfolio	α	DP_m	RI	r_f	β_m	$DP_m \times r_m$	$RI \times r_m$	$r_f \times r_m$	$\bar{R}^2$
				Panel A: Size-sorted portfolios					
Low	16.903***	−0.021***	0.057***	2.635***	−0.513***	0.027***	−0.093***	2.180***	0.418
2	9.651**	−0.014*	0.034**	2.568***	−0.734***	0.053***	−0.109***	4.154***	0.472
3	6.127	−0.006	0.047***	1.743***	−0.773***	0.068***	−0.108***	3.295***	0.503
4	3.000	−0.005	0.033**	2.111***	−0.767***	0.073***	−0.070***	1.488***	0.509
5	1.374	−0.001	0.029**	1.458**	−0.808***	0.102***	−0.089***	−0.439	0.530
6	1.831	−0.002	0.033**	1.519***	−0.748***	0.100***	−0.078***	−2.574***	0.544
7	0.216	0.004	0.032**	0.910*	−0.702***	0.074***	−0.110***	−1.365***	0.552
8	−3.382	0.009	0.030**	0.620	−0.734***	0.080***	−0.100***	−1.592***	0.572
9	−9.029***	0.019***	0.034***	−0.378	−0.713***	0.072***	−0.135***	−1.452***	0.598
High	−2.870	0.011**	0.001	−0.875*	−0.728***	0.093***	−0.111***	−5.711***	0.602

Panel B: Book-to-market sorted portfolios									
Low	−6.283*	0.014**	0.016	0.111	−0.757***	0.117***	−0.116***	−10.546***	0.583
2	−5.904*	0.017***	0.019	−0.517	−0.683***	0.068***	−0.059***	−4.718***	0.572
3	−5.827*	0.011**	0.001	0.447	−0.616***	0.058***	−0.087***	−4.205***	0.553
4	−7.678**	0.019***	0.026**	−0.465	−0.621***	0.028***	−0.129***	0.930**	0.555
5	−6.161*	0.014**	−0.004	−0.278	−0.621***	0.029***	−0.144***	2.078***	0.533
6	−3.422	0.009*	0.012	0.044	−0.640***	0.028***	−0.091***	3.606***	0.531
7	−5.408	0.012**	0.016	0.045	−0.564***	0.001	−0.090***	4.727***	0.512
8	−4.225	0.011*	0.013	−0.081	−0.541***	−0.033***	−0.201***	9.742***	0.506
9	−1.528	0.010*	0.013	−0.361	−0.610***	−0.011*	−0.092***	7.786***	0.502
High	2.165	0.008	0.025	−0.287	−0.606***	−0.041***	−0.111***	10.222***	0.480
Panel C: Momentum-sorted portfolios									
Low	22.801***	−0.022**	0.026	2.358**	−1.060***	0.146***	−0.464***	−0.214	0.492
2	7.756*	−0.004	0.043**	0.941	−0.863***	0.095***	−0.378***	2.247***	0.537
3	1.128	0.007	0.057***	−0.017	−0.693***	0.055***	−0.279***	2.059***	0.542
4	−1.737	0.014**	0.040***	−0.850*	−0.638***	0.037***	−0.241***	1.553***	0.557
5	−7.023**	0.013**	0.007	0.350	−0.617***	0.034***	−0.155***	0.965**	0.540
6	−11.497***	0.020***	0.010	−0.001	−0.562***	0.016***	−0.091***	0.495	0.527
7	−6.198*	0.010*	0.005	0.424	−0.571***	0.019***	−0.060***	−0.452	0.536
8	−10.694***	0.018***	0.008	−0.051	−0.587***	0.025***	−0.004	−1.139**	0.514
9	−8.405**	0.013**	0.013	0.337	−0.664***	0.043***	−0.003	−2.825***	0.529
High	−9.361*	0.021***	0.013	−0.237	−0.901***	0.114***	0.096***	−8.420***	0.488

continued

Table 3.4 Continued

Portfolio	α	DP_m	RI	r_f	β_m	$DP_m \times r_m$	$RI \times r_m$	$r_f \times r_m$	$\bar{R}^2$
			Panel D: Short-term reversal sorted portfolios						
Low	−4.559	0.005	0.008	−0.603	−1.239***	0.244***	−0.405***	−6.598***	0.434
2	−6.485	0.018**	0.019	−0.960	−0.993***	0.144***	−0.247***	−3.907***	0.518
3	−8.111*	0.015**	−0.007	0.251	−0.768***	0.069***	−0.161***	−1.891***	0.520
4	−4.821	0.011*	0.017	0.360	−0.736***	0.081***	−0.137***	−3.348***	0.544
5	−4.391	0.012**	0.006	0.064	−0.707***	0.062***	−0.131***	−0.668	0.563
6	−2.381	0.006	0.004	0.534	−0.635***	0.044***	−0.120***	−0.307	0.549
7	−8.269**	0.016***	0.013	0.057	−0.626***	0.045***	−0.116***	−0.781*	0.566
8	−1.405	0.005	0.004	0.738	−0.622***	0.045***	−0.064***	−1.078**	0.552
9	−4.072	0.008	−0.005	0.736	−0.589***	0.022***	−0.087***	0.722	0.535
High	14.158***	−0.014**	−0.005	2.172***	−0.763***	0.051***	−0.109***	0.244	0.531
			Panel E: Long-term reversal sorted portfolios						
Low	7.009	0.005	0.010	−0.024	−0.790***	−0.005	0.025*	7.020***	0.491
2	−2.562	0.015**	0.012	−0.620	−0.667***	0.003	−0.004	3.847***	0.514
3	−8.508**	0.020***	0.018	−0.615	−0.561***	−0.008	−0.052***	3.329***	0.509
4	−9.008***	0.018***	0.034**	−0.560	−0.546***	−0.012**	−0.105***	4.522***	0.514
5	−8.695**	0.018***	0.030**	−0.392	−0.534***	−0.009*	−0.100***	3.367***	0.515
6	−10.796***	0.021***	0.003	−0.391	−0.508***	0.003	−0.083***	1.024**	0.515
7	−9.079***	0.017***	0.015	−0.162	−0.523***	0.017***	−0.110***	−0.473	0.523
8	−5.942*	0.017***	0.021*	−0.660	−0.614***	0.044***	−0.091***	−2.623***	0.546
9	−11.601***	0.025***	0.032**	−0.658	−0.628***	0.030***	−0.161***	−1.218**	0.547
High	−7.218	0.017**	0.013	0.097	−0.914***	0.142***	−0.242***	−8.859***	0.566

Panel F: Volatility-sorted portfolios									
High	18.237***	−0.010	0.044**	0.760	−0.540***	−0.005	−0.105***	2.210***	0.356
2	14.296***	−0.007	0.065***	1.215*	−0.651***	0.008	−0.160***	2.103***	0.470
3	10.439**	−0.005	0.062***	1.355**	−0.614***	−0.008	−0.148***	3.275***	0.489
4	4.704	0.001	0.055***	0.969	−0.592***	−0.002	−0.174***	4.216***	0.500
5	1.438	0.006	0.053***	0.761	−0.529***	−0.013**	−0.163***	4.846***	0.502
6	−0.984	0.006	0.044***	0.881	−0.506***	−0.001	−0.175***	4.027***	0.510
7	1.927	0.002	0.041***	0.748	−0.437***	0.002	−0.140***	2.110***	0.498
8	2.264	0.006	0.041***	0.121	−0.347***	−0.010**	−0.118***	1.746***	0.481
9	4.710*	0.004	0.046***	−0.228	−0.224***	−0.018***	−0.125***	0.526	0.433
Low	4.604**	0.002	0.040***	−0.129	−0.045***	−0.033***	−0.058***	−1.301***	0.270
Panel G: Industry-sorted portfolios									
NoDur	−4.783	0.014***	0.020	−0.244	−0.334***	−0.034***	−0.030***	−0.092	0.440
Durbl	2.941	−0.000	0.019	0.710	−0.715***	−0.004	−0.101***	5.094***	0.474
Manuf	−7.689**	0.021***	0.045***	−0.531	−0.561***	−0.005	−0.107***	1.045**	0.548
Enrgy	−17.546***	0.029***	0.016	−0.528	−0.451***	−0.060***	−0.138***	5.935***	0.349
HiTec	9.441**	−0.010	0.001	0.989	−1.157***	0.256***	−0.078***	−16.612***	0.529
Telcm	2.089	0.005	−0.010	−0.645	−0.807***	0.102***	−0.083***	1.554***	0.437
Shops	−11.652***	0.025***	0.013	−0.243	−0.576***	0.033***	−0.030***	−3.217***	0.473
Hlth	−9.522**	0.031***	−0.013	−1.721***	−0.404***	−0.033***	−0.018	−1.722***	0.420
Utils	−4.346	0.013**	−0.005	0.181	−0.349***	−0.024***	−0.113***	5.830***	0.320
Other	−9.456**	0.023***	0.021	−0.241	−0.793***	0.065***	−0.186***	1.906***	0.556

Chapter 4

Performance Sensitivity

4.1 Alternative Set-Ups

In this section, I report the results of several robustness checks performed on the previously reported empirical findings. First, I show evidence of the MA strategy performance in two subperiods of equal length. Second, I show how the MA strategy performs when various lag lengths are used. Third, I report the intensity of trading, the BETC, the probability of being on the right side of the market, and the statistical significance of the mean return and standard deviation improvement. Finally, I also report how the number of trades and the BETC vary with alternative lengths of the MA.[1]

4.1.1 Subperiods

In this subsection, I split the sample in two when the first half-period starts in January 1960 and ends in December 1986 while the second half-period starts in January 1987 and ends in December 2011. Table 4.1 presents the results for the risk-adjusted abnormal MAP returns using the single-factor CAPM model, the three-factor Fama-French model, as well as the four-factor Carhart model. In the interest of preserving space and focus the reader's attention on the important results, only the alphas and the adjusted R^2 are reported in Table 4.1.[2] Overall, the results in the previous section are robust with respect to the two subperiods. The results are a little weaker with respect to the alphas of size deciles (Panel A) and momentum deciles (Panel C) where half of the decile portfolios' alphas are insignificant in the first half-period while all ten deciles' alphas are highly significant in the second half-period.

4.1.2 *Alternative Lag Lengths*

Next, I investigate the effect of MA windows of various lengths on the size of the average MAP returns for all the sets of portfolios under investigation. Table 4.2 reports the annualized average returns of the MAP for MA windows of 5 days, 10 days, 60 days, 120 days, and 250 days in length. The average returns are economically and statistically significant with fewer as well as more than 20 days, the baseline window used previously. These results persist with up to 60 days in the MA length and appear to be much weaker at longer windows of 120 and 250 months. Importantly, significant cross-sectional variation persists for all sets of portfolios with the exception of book-to-market portfolios. The range of annual MAP returns with an MA window of 5 days is between roughly −2% and 21%. The range of annual MAP returns with the length of the MA of 10 days is between approximately −5% and 18%. When I increase the MA window length to 60 days the range of average annualized MAP returns drops considerably to between −8% and 12%, depending on which sets of deciles I consider.

4.1.3 *Statistical Significance, Trading Intensity, and BETC*

Table 4.3 reports the annualized change in the average return between the MA and BH portfolio, the change in the annualized standard deviation, the proportion of times when the signal is on, p_A, the number of one-way transactions, NT, the BETC that would eliminate the return improvement based on the number of transactions NT, and two success probabilities, p_1 and p_2. The first probability reports the fraction of times when a signal resulted in an action that led to a positive return, while the second probability reports the fraction of times when the return was in excess of the risk-free rate.

Table 4.3 reports the statistical significance in the improvement of the average return $\Delta\mu$ of the MA portfolio over the BH

Table 4.1 Factor regressions results in subperiods

This table reports alphas and adjusted R^2 of the regressions of the MAP excess returns on the market factor, the Fama-French three-factors, and the Carhart four-factors in two subperiods using portfolios sorted by various characteristics. The alphas are annualized and in percent. The sample period covers January 4, 1960 until December 31, 2013 with value-weighted portfolio returns. The length of the MA window is 20 days. Newey and West (1987) standard errors with 5 lags are used in reporting statistical significance of a two-sided null hypothesis at the 1%, 5%, and 10% level given by a***, a**, and a*, respectively.

Portfolio	α	$\bar{R}^2$	α	$\bar{R}^2$	α	$\bar{R}^2$	α	$\bar{R}^2$	α	$\bar{R}^2$	α	$\bar{R}^2$
	1960/01/04–1987/12/31						1988/01/04–2013/12/31					
	CAPM		Fama-French		Carhart		CAPM		Fama-French		Carhart	
						Panel A: Size-sorted portfolios						
Low	19.882***	0.386	22.014***	0.614	22.683***	0.618	12.930***	0.420	14.135***	0.538	15.053***	0.546
2	17.917***	0.440	19.738***	0.618	20.724***	0.626	8.770***	0.465	10.195***	0.576	11.143***	0.581
3	16.159***	0.455	17.923***	0.618	18.940***	0.627	7.715***	0.503	8.932***	0.593	9.762***	0.597
4	16.661***	0.482	18.145***	0.620	19.375***	0.633	5.799***	0.506	6.787***	0.580	7.638***	0.585
5	14.443***	0.489	15.829***	0.614	17.036***	0.627	5.047**	0.528	5.741***	0.583	6.342***	0.585
6	14.465***	0.517	15.694***	0.602	16.971***	0.617	4.538**	0.539	4.905***	0.565	5.472***	0.567
7	14.287***	0.542	15.289***	0.606	16.436***	0.617	4.289**	0.542	4.724***	0.557	5.431***	0.560
8	13.050***	0.563	14.153***	0.598	15.374***	0.611	2.790	0.562	3.112*	0.568	3.707**	0.570
9	10.697***	0.581	11.391***	0.592	12.595***	0.604	0.376	0.589	0.635	0.590	1.242	0.593
High	5.426***	0.587	4.978***	0.594	6.068***	0.602	−0.515	0.595	−0.973	0.605	−0.870	0.605

continued

Table 4.1 Continued

Portfolio	α	$\overline{R}^2$	α	$\overline{R}^2$	α	$\overline{R}^2$	α	$\overline{R}^2$	α	$\overline{R}^2$	α	$\overline{R}^2$
	1960/01/04–1987/12/31						1988/01/04–2013/12/31					
	CAPM		Fama-French		Carhart		CAPM		Fama-French		Carhart	
					Panel B: Book-to-market sorted portfolios							
Low	11.475***	0.584	9.310***	0.600	10.326***	0.606	2.163	0.558	0.958	0.591	0.986	0.591
2	10.261***	0.578	9.651***	0.584	10.776***	0.593	0.542	0.561	0.122	0.566	0.579	0.567
3	10.504***	0.562	10.461***	0.566	11.691***	0.577	0.365	0.541	0.179	0.544	0.998	0.549
4	10.736***	0.561	11.139***	0.564	12.174***	0.573	1.132	0.541	1.702	0.553	2.270	0.556
5	8.471***	0.535	8.881***	0.539	10.199***	0.555	−1.401	0.515	−0.600	0.538	−0.314	0.539
6	7.718***	0.511	8.698***	0.525	9.733***	0.535	1.515	0.524	2.296	0.542	2.698	0.543
7	8.095***	0.499	9.628***	0.513	10.852***	0.528	0.742	0.504	1.751	0.538	2.390	0.541
8	7.512***	0.514	9.878***	0.545	10.990***	0.557	0.947	0.452	2.785	0.537	2.772	0.537
9	8.301***	0.503	10.910***	0.538	12.292***	0.555	1.116	0.476	2.612	0.535	3.350*	0.539
High	9.773***	0.491	12.533***	0.536	13.963***	0.551	4.196*	0.446	6.147***	0.518	6.834***	0.520

Panel C: Momentum-sorted portfolios

Low	22.250***	0.526	22.470***	0.550	20.399***	0.570	20.063***	0.435	21.333***	0.450	15.932***	0.516
2	15.459***	0.529	15.466***	0.536	13.954***	0.552	9.937***	0.490	10.602***	0.496	6.827***	0.548
3	11.878***	0.537	11.945***	0.541	11.010***	0.548	6.440***	0.505	6.978***	0.516	4.683**	0.544
4	10.423***	0.561	10.492***	0.562	10.541***	0.562	3.752**	0.523	4.281**	0.535	3.220*	0.542
5	10.327***	0.552	10.639***	0.556	11.211***	0.559	−0.357	0.518	−0.077	0.524	−0.386	0.524
6	10.492***	0.549	10.678***	0.549	11.711***	0.558	−1.756	0.509	−1.487	0.515	−1.122	0.516
7	9.152***	0.562	9.668***	0.565	11.194***	0.583	−0.820	0.519	−0.598	0.524	0.546	0.534
8	8.240***	0.537	9.074***	0.542	11.351***	0.580	−2.560	0.502	−2.356	0.506	−0.494	0.534
9	9.049***	0.527	9.697***	0.538	12.669***	0.595	−0.932	0.529	−0.821	0.530	1.603	0.569
High	11.461***	0.493	12.016***	0.517	16.166***	0.594	0.503	0.464	−0.039	0.487	4.017*	0.553

continued

Table 4.1 Continued

Portfolio	α	$\bar{R}^2$	α	$\bar{R}^2$	α	$\bar{R}^2$	α	$\bar{R}^2$	α	$\bar{R}^2$	α	$\bar{R}^2$
	1960/01/04–1987/12/31						1988/01/04–2013/12/31					
	CAPM		Fama-French		Carhart		CAPM		Fama-French		Carhart	
					Panel D: Short-term reversal sorted portfolios							
Low	−2.507	0.372	−3.574	0.381	−1.871	0.392	1.031	0.409	1.542	0.417	−0.249	0.423
2	7.093***	0.491	6.946***	0.496	8.377***	0.505	2.059	0.502	2.296	0.503	1.729	0.504
3	9.793***	0.540	9.800***	0.544	11.139***	0.554	0.007	0.497	0.278	0.499	0.565	0.499
4	11.070***	0.553	11.497***	0.556	12.705***	0.565	1.938	0.525	2.048	0.526	2.606	0.528
5	10.202***	0.559	10.526***	0.563	11.637***	0.573	1.277	0.550	1.431	0.552	1.958	0.554
6	8.930***	0.543	9.369***	0.552	10.497***	0.562	1.680	0.539	1.823	0.544	2.411	0.546
7	10.524***	0.569	11.038***	0.574	12.000***	0.582	−0.320	0.554	−0.141	0.557	0.308	0.558
8	11.584***	0.562	12.102***	0.568	13.141***	0.577	0.679	0.540	0.770	0.541	1.618	0.547
9	11.872***	0.550	12.726***	0.562	13.675***	0.570	−0.870	0.520	−0.642	0.522	−0.478	0.522
High	18.201***	0.524	19.173***	0.551	20.198***	0.558	10.962***	0.528	11.165***	0.532	10.639***	0.533

Panel E: Long-term reversal sorted portfolios

Low	13.123***	0.481	15.138***	0.509	16.686***	0.521	8.129***	0.488	9.460***	0.532	10.740***	0.539
2	9.604***	0.512	11.261***	0.529	12.570***	0.540	1.979	0.512	2.865	0.530	4.131**	0.540
3	7.673***	0.511	8.856***	0.519	10.042***	0.530	0.565	0.502	1.345	0.518	2.459	0.527
4	7.040***	0.514	8.063***	0.519	9.163***	0.528	−0.878	0.499	0.034	0.524	0.552	0.526
5	8.741***	0.534	9.638***	0.538	10.643***	0.546	−0.138	0.495	0.752	0.521	1.337	0.524
6	8.623***	0.540	9.067***	0.541	10.283***	0.554	−2.026	0.497	−1.633	0.513	−0.980	0.517
7	8.284***	0.537	8.209***	0.539	9.402***	0.551	−1.098	0.509	−0.840	0.519	−0.239	0.523
8	8.476***	0.547	8.126***	0.550	9.280***	0.560	0.539	0.541	0.574	0.549	1.547	0.556
9	12.001***	0.563	11.395***	0.570	12.237***	0.574	−0.323	0.526	−0.477	0.533	0.401	0.537
High	14.837***	0.563	13.379***	0.590	14.502***	0.596	1.815	0.536	0.796	0.552	0.606	0.552

continued

Table 4.1 Continued

Portfolio	α	$\bar{R}^2$	α	$\bar{R}^2$	α	$\bar{R}^2$	α	$\bar{R}^2$	α	$\bar{R}^2$	α	$\bar{R}^2$
	1960/01/04–1987/12/31						**1988/01/04–2013/12/31**					
	CAPM		**Fama-French**		**Carhart**		**CAPM**		**Fama-French**		**Carhart**	
Panel F: Volatility-sorted portfolios												
High	20.010***	0.375	22.181***	0.534	22.830***	0.536	11.931***	0.348	13.321***	0.415	13.708***	0.416
2	21.383***	0.454	23.307***	0.596	24.289***	0.600	13.420***	0.475	14.728***	0.526	14.844***	0.526
3	20.415***	0.474	22.430***	0.608	23.690***	0.616	10.955***	0.494	12.265***	0.542	12.509***	0.542
4	17.571***	0.482	19.448***	0.610	20.433***	0.616	7.462***	0.493	8.740***	0.536	9.035***	0.537
5	16.379***	0.488	18.173***	0.608	19.290***	0.617	6.283***	0.488	7.508***	0.530	7.750***	0.530
6	14.549***	0.498	16.081***	0.593	17.130***	0.603	4.456**	0.489	5.564***	0.526	5.841***	0.526
7	13.053***	0.506	14.509***	0.585	15.366***	0.593	4.701***	0.476	5.491***	0.501	5.943***	0.504
8	12.003***	0.511	13.361***	0.574	14.155***	0.582	5.368***	0.450	6.017***	0.475	6.444***	0.478
9	10.133***	0.503	11.394***	0.555	11.925***	0.560	6.096***	0.367	6.685***	0.401	6.989***	0.403
Low	9.027***	0.461	10.191***	0.510	10.508***	0.513	4.974***	0.128	5.173***	0.143	5.436***	0.146

Panel G: Industry-sorted portfolios

NoDur	10.771***	0.542	11.553***	0.551	12.670***	0.562	−0.150	0.376	−0.291	0.390	0.519	0.397
Durbl	8.002***	0.490	8.029***	0.490	9.141***	0.497	5.579**	0.454	6.662**	0.472	6.962***	0.472
Manuf	11.882***	0.595	12.393***	0.602	13.595***	0.612	2.393	0.516	2.748	0.521	3.532*	0.525
Enrgy	8.440***	0.427	8.661***	0.432	9.510***	0.436	−3.877	0.296	−3.262	0.312	−1.533	0.325
HiTec	8.839***	0.508	6.548***	0.521	7.718***	0.527	8.857***	0.484	6.956***	0.542	6.139**	0.544
Telcm	4.244***	0.335	5.581***	0.358	6.171***	0.361	2.395	0.462	2.344	0.469	2.219	0.469
Shops	14.014***	0.529	14.119***	0.541	15.083***	0.547	−0.322	0.435	−0.454	0.436	0.060	0.438
Hlth	10.842***	0.523	10.395***	0.525	11.394***	0.530	−0.867	0.355	−1.453	0.367	−0.649	0.371
Utils	11.198***	0.414	12.522***	0.431	12.892***	0.433	0.242	0.260	0.967	0.308	2.037	0.318
Other	13.941***	0.571	14.980***	0.609	16.255***	0.621	2.438	0.528	3.599*	0.554	3.355	0.554

Table 4.2 Alternative MA lag lengths.
This table reports the sensitivity of the average annualized MAP(q) excess returns to the length of the MA using portfolios sorted by various characteristics. The sample period covers January 4, 1960 until December 31, 2013 with value-weighted portfolio returns. The length of the MA window is q days. Newey and West (1987) standard errors with q lags are used in adjusting the standard errors of the parameter estimates.

Portfolio	MAP(5)	MAP(10)	MAP(60)	MAP(120)	MAP(250)
		Panel A: Size-sorted portfolios			
Low	17.81	15.97	10.15	6.88	3.76
2	13.51	11.96	7.08	3.72	1.63
3	12.32	10.20	5.31	2.53	0.38
4	11.29	9.36	4.55	2.27	0.71
5	11.19	8.08	3.89	2.01	0.66
6	11.42	9.19	3.73	2.13	0.35
7	11.00	7.47	3.63	1.77	0.86
8	9.35	6.80	2.40	1.02	0.58
9	6.52	3.77	1.69	0.65	0.03
High	0.92	−0.49	−1.49	−0.87	0.24
		Panel B: Book-to-market sorted portfolios			
Low	5.28	3.14	1.25	1.37	1.05
2	4.82	2.82	0.47	−0.55	−0.90
3	4.83	2.65	0.80	−0.26	0.48
4	5.85	3.95	1.52	0.36	−0.14
5	4.73	1.17	0.99	0.35	0.15
6	5.22	3.45	0.35	−0.71	−0.80
7	4.87	1.12	1.16	0.73	−0.74
8	4.87	1.93	−0.32	−1.12	−1.07
9	4.53	2.91	0.71	−0.50	−0.89
High	6.92	6.12	2.66	1.75	−0.54
		Panel C: Momentum-sorted portfolios			
Low	20.60	17.73	10.29	9.41	5.38
2	11.97	9.52	4.81	3.50	1.25
3	6.54	5.14	1.88	1.66	0.54
4	5.83	3.87	0.51	1.12	0.32
5	4.12	2.63	0.21	0.66	0.56
6	3.55	2.29	0.18	0.29	−0.75
7	4.07	3.02	0.11	0.18	0.86
8	2.17	−0.12	−1.00	−0.86	−0.54
9	4.68	1.37	0.85	−0.22	0.61
High	7.98	2.57	0.85	−0.88	−0.67

continued

Table 4.2 Continued

Portfolio	MAP(5)	MAP(10)	MAP(60)	MAP(120)	MAP(250)
		Panel D: Short-term reversal sorted portfolios			
Low	−2.31	−4.91	−8.40	−7.13	−6.35
2	4.54	0.78	−3.31	−2.42	−3.02
3	5.41	2.75	−0.50	−1.59	−2.82
4	5.92	4.57	0.78	−0.10	−1.41
5	3.84	2.59	0.86	0.32	−0.56
6	3.85	1.97	1.10	0.12	−0.60
7	3.12	2.39	1.67	1.10	1.79
8	4.82	3.77	2.85	2.39	1.77
9	5.07	3.46	2.10	2.73	1.99
High	14.43	12.29	12.23	12.89	10.87
		Panel E: Long-term reversal sorted portfolios			
Low	8.29	8.09	2.14	1.21	−1.25
2	5.32	3.91	0.71	−0.03	−1.34
3	4.22	1.73	−0.18	−0.47	−1.41
4	3.13	−0.05	−0.37	−0.82	−0.29
5	3.11	1.76	0.38	−0.64	−0.06
6	2.08	0.65	−0.34	−0.89	−0.22
7	3.08	1.46	−0.08	−0.39	−0.43
8	3.53	2.08	0.16	0.15	−0.59
9	6.42	3.73	1.31	1.27	1.03
High	9.12	5.82	3.29	2.37	1.17
		Panel F: Volatility-sorted portfolios			
High	15.66	14.92	7.51	2.80	0.56
2	19.71	16.22	7.58	3.95	1.78
3	16.87	14.49	6.88	4.46	2.31
4	13.51	10.95	5.40	2.79	1.31
5	12.43	10.21	4.23	2.36	1.18
6	12.38	9.75	3.54	1.83	0.44
7	10.53	8.28	3.43	1.92	0.80
8	9.93	7.45	3.45	1.78	0.74
9	9.24	7.71	3.56	2.01	0.55
Low	9.62	7.98	3.38	2.12	1.44

continued

Table 4.2 Continued

Portfolio	MAP(5)	MAP(10)	MAP(60)	MAP(120)	MAP(250)
		Panel G: Industry-sorted portfolios			
Nodur	4.75	3.41	0.94	0.59	−0.01
Durbl	7.69	4.86	3.22	2.40	0.73
Manuf	7.17	4.87	2.14	0.78	−0.02
Enrgy	2.58	−0.68	−2.77	−3.16	−2.80
HiTec	5.05	3.96	1.05	0.85	0.20
Telcm	1.21	−0.03	−0.49	0.01	0.16
Shops	7.39	6.09	1.76	0.34	−1.15
Hlth	5.47	3.24	−0.28	−0.53	−1.18
Utils	7.52	6.09	1.97	0.74	1.02
Other	10.38	6.58	2.84	1.45	0.76

Table 4.3 Trading frequency and BETC

This table reports the results for the improvement delivered by the MA switching strategy over the buy-and-hold strategy, the trading frequency, as well as the BETC using ten decile portfolios sorted by various characteristics. The sample period covers January 4, 1960 until December 31, 2013 with value-weighted portfolio returns. $\Delta\mu$ is the annualized improvement in the average in-sample monthly return, $\Delta\sigma$ is the annualized improvement in the return standard deviation, p_A is the proportion of months during which there is a hold signal, NT is the number of transactions (buy or sell) over the entire sample period, BETC is the break-even one-sided transaction cost in percent, p_1 is the proportion of months during which a buy signal was followed by a positive return of the underlying portfolios, and p_2 is the proportion of months during which a buy signal was followed by a portfolio return in excess of the risk-free rate. The length of the MA window is 20 days.

Portfolio	$\Delta\mu$	$\Delta\sigma$	p_A	NT	BETC	p_1	p_2
			Panel A: Size-sorted portfolios				
Low	13.89	5.24	0.60	931	0.81	76.42	57.78
2	10.10	6.19	0.61	1071	0.51	74.19	56.44
3	8.60	6.14	0.61	1089	0.43	73.39	56.48
4	7.92	6.02	0.62	1147	0.38	73.09	56.08
5	6.38	5.92	0.62	1139	0.30	72.44	56.05
6	6.27	5.60	0.62	1165	0.29	72.64	55.84
7	5.96	5.67	0.62	1145	0.28	72.39	55.50
8	4.50	5.71	0.62	1193	0.20	71.24	54.99
9	2.14	5.72	0.62	1313	0.09	70.94	54.43
High	−1.22	5.86	0.61	1501	−0.04	69.14	52.85

continued

Table 4.3 Continued

Portfolio	$\Delta\mu$	$\Delta\sigma$	p_A	NT	BETC	p_1	p_2
			Panel B: Book-to-market sorted portfolios				
Low	2.93	6.60	0.59	1409	0.11	70.88	52.84
2	1.90	5.83	0.61	1339	0.08	70.29	53.19
3	2.22	5.49	0.61	1345	0.09	70.12	53.07
4	2.57	5.59	0.61	1315	0.11	70.54	53.69
5	0.28	5.52	0.61	1407	0.01	69.44	53.15
6	1.47	5.33	0.62	1259	0.06	69.84	53.77
7	1.36	5.14	0.62	1301	0.06	70.12	53.95
8	1.10	5.55	0.63	1315	0.05	69.99	54.23
9	1.60	5.51	0.63	1309	0.07	69.98	54.11
High	3.57	6.05	0.62	1281	0.15	70.32	53.97
			Panel C: Momentum-sorted portfolios				
Low	16.15	9.08	0.52	1197	0.73	73.87	50.78
2	8.54	7.08	0.56	1295	0.36	71.94	51.25
3	5.58	5.91	0.58	1315	0.23	71.41	52.40
4	3.63	5.76	0.59	1359	0.14	70.77	52.55
5	1.81	5.42	0.60	1361	0.07	70.46	52.98
6	1.28	5.33	0.62	1401	0.05	70.16	53.48
7	0.96	5.41	0.62	1355	0.04	69.83	53.58
8	−0.32	5.37	0.63	1389	−0.01	69.72	54.65
9	0.59	6.06	0.62	1435	0.02	69.95	54.20
High	1.95	7.73	0.64	1307	0.08	70.93	55.74
			Panel D: Short-term reversal sorted portfolios				
Low	−5.84	10.44	0.72	1241	−0.26	69.11	57.69
2	0.09	8.68	0.64	1277	0.00	70.18	54.68
3	1.05	7.40	0.62	1345	0.04	70.21	53.64
4	2.88	6.57	0.61	1299	0.12	70.55	53.37
5	2.20	6.09	0.61	1355	0.09	69.98	53.45
6	2.02	5.49	0.62	1315	0.08	70.07	53.56
7	1.92	5.23	0.60	1413	0.07	70.43	53.13
8	3.08	4.87	0.58	1401	0.12	70.84	52.97
9	2.43	4.82	0.57	1409	0.09	71.58	53.08
High	10.74	6.28	0.47	1475	0.40	74.41	50.48

continued

Table 4.3 Continued

Portfolio	$\Delta\mu$	$\Delta\sigma$	p_A	NT	BETC	p_1	p_2
Panel E: Long-term reversal sorted portfolios							
Low	6.60	6.97	0.58	1295	0.28	71.55	53.33
2	2.23	5.96	0.61	1309	0.09	70.62	53.92
3	0.94	5.33	0.62	1429	0.04	69.70	53.53
4	0.03	5.30	0.62	1389	0.00	69.50	53.57
5	1.11	5.25	0.62	1363	0.04	69.73	53.53
6	0.33	4.98	0.63	1311	0.01	69.20	53.51
7	0.55	5.07	0.62	1407	0.02	69.79	53.95
8	1.17	5.66	0.61	1397	0.05	70.17	53.05
9	2.30	6.30	0.60	1425	0.09	70.19	52.84
High	3.89	7.93	0.59	1375	0.15	71.13	53.36
Panel F: Volatility-sorted portfolios							
High	12.84	6.21	0.65	935	0.75	73.77	57.91
2	13.59	7.06	0.61	1097	0.67	74.03	56.13
3	12.06	6.76	0.61	1041	0.63	74.05	56.20
4	9.18	6.18	0.62	1053	0.47	73.82	56.70
5	8.27	5.66	0.63	1025	0.44	73.53	57.10
6	6.63	5.23	0.64	1047	0.34	72.74	57.05
7	6.35	4.71	0.65	1091	0.32	73.20	57.61
8	6.50	4.24	0.66	989	0.36	72.62	57.72
9	6.37	3.65	0.67	927	0.37	73.23	58.60
Low	6.05	2.75	0.68	831	0.40	74.62	59.99
Panel G: Industry-sorted portfolios							
NoDur	2.85	4.59	0.63	1299	0.12	70.85	54.62
Durbl	2.59	6.89	0.57	1391	0.10	70.41	51.15
Manuf	3.75	5.81	0.61	1315	0.15	70.42	53.22
Enrgy	−0.93	6.54	0.59	1431	−0.04	69.90	52.46
HiTec	4.16	8.28	0.57	1407	0.16	71.72	52.39
Telcm	0.05	5.87	0.58	1461	0.00	70.00	51.28
Shops	3.64	5.70	0.60	1313	0.15	71.20	53.47
Hlth	2.03	5.73	0.60	1331	0.08	70.88	53.13
Utils	3.84	4.60	0.61	1213	0.17	71.47	53.87
Other	4.47	6.55	0.61	1273	0.19	71.02	53.94

portfolio as well as the reduction in the return standard deviation $\Delta\sigma$. The evidence points toward a substantial improvement in a mean–variance sense for all sets of portfolios under consideration. The annualized improvement in the average return ranges from −1% to 16% while the reduction in the standard deviation is between approximately 3% and 10%. The MA strategy is active more often than not ranging between 52% and 72% of the sample. Yet, the number of transactions, NT, is never above 1,500 and can be as little as 800 for lowest total volatility decile. In a sample that exceeds 13,000 days this translates into average holding periods of between 10 and 22 days where the MA strategy is continuously invested either in the risky asset or the risk-free asset. Next, I calculate BETC as the level of one-way proportional transaction cost in percent that would eliminate completely the average MAP portfolio return. The values of the BETC for the various sets of portfolios range between −0.04% and as high as 0.81%. This is a very large level of transaction cost which should more than compensate for the rebalancing costs associated with implementing the value-weighted portfolio scheme used to construct the portfolio returns at least for large institutional investors. Finally, the last two columns report the fraction of months that the MA strategy generates a positive return (p_1) as well as a return that is in excess of the risk-free rate (p_2). I report the statistical significance of the null hypothesis that the true fraction of times is above 50%. With the exception of three momentum deciles and two industry portfolios, all the observed fractions are highly statistically significant and range from 50.5% to 76.4% success rate of the MA strategy being on the right side of the market. These are considerably favorable odds and in line with the evidence reported previously about the superior performance of the MA switching strategy.

Table 4.4 reports the number of transactions and the BETC for alternative lag lengths of the MA strategy. As would be expected, the trading intensity declines as I implement the MA strategy at longer window lengths of up to 250 days. Vice versa, at shorter MA window length of 5 and 10 days the number of transaction increases but it is never excessive with a maximum of 3,400 transaction for MAP(5) and 2,300 for MAP(10). In a similar fashion, the BETC decline at shorter MA window lengths compared to the baseline MAP(20) results while they increase at longer MA window lengths.

The large values of BETC and the relatively small number of transactions NT relative to the total sample size suggest that this

Table 4.4 Trading and BETC at various MA lags
This table reports the number of trades and the BETC involved in the MA switching strategy over the buy-and-hold strategy using ten decile portfolios sorted by various characteristics. The sample period covers January 4, 1960 until December 31, 2013 with value-weighted portfolio returns. NT is the number of transactions (buy or sell) over the entire sample period and BETC is the break-even one-sided transaction cost in percent.

Portfolio	MAP(5)		MAP(10)		MAP(60)		MAP(120)		MAP(250)	
	NT	BETC	NT	BETC	NT	BETC	NT	BETC	NT	BETC
				Panel A: Size-sorted portfolios						
Low	2456	0.39	1484	0.58	380	1.45	248	1.49	148	1.36
2	2684	0.27	1714	0.38	467	0.82	334	0.60	226	0.38
3	2690	0.25	1726	0.32	503	0.57	334	0.41	234	0.09
4	2787	0.22	1778	0.29	531	0.46	377	0.32	226	0.17
5	2769	0.22	1808	0.24	601	0.35	386	0.28	216	0.16
6	2767	0.22	1742	0.29	563	0.36	382	0.30	233	0.08
7	2795	0.21	1809	0.22	577	0.34	388	0.25	218	0.21
8	2853	0.18	1831	0.20	605	0.22	430	0.13	224	0.14
9	2965	0.12	1938	0.11	654	0.14	430	0.08	230	0.01
High	3237	0.02	2115	−0.01	785	−0.10	530	−0.09	246	0.05
				Panel B: Book-to-market sorted portfolios						
Low	3181	0.09	2105	0.08	735	0.09	472	0.16	264	0.21
2	3091	0.08	2032	0.08	717	0.04	516	−0.06	315	−0.15
3	3153	0.08	2069	0.07	683	0.06	489	−0.03	236	0.11
4	3095	0.10	1956	0.11	623	0.13	470	0.04	274	−0.03
5	3094	0.08	2070	0.03	657	0.08	438	0.04	220	0.04
6	3045	0.09	1947	0.10	667	0.03	496	−0.08	272	−0.16
7	3063	0.09	2016	0.03	653	0.10	414	0.10	254	−0.16
8	3099	0.09	2011	0.05	695	−0.02	478	−0.13	244	−0.23
9	3055	0.08	1949	0.08	617	0.06	396	−0.07	206	−0.23
High	3061	0.12	1896	0.18	599	0.24	360	0.26	260	−0.11
				Panel C: Momentum-sorted portfolios						
Low	2953	0.38	1831	0.53	643	0.87	380	1.33	309	0.93
2	3039	0.21	1931	0.27	645	0.40	452	0.42	256	0.26
3	3145	0.11	2060	0.14	677	0.15	452	0.20	274	0.10
4	3092	0.10	2063	0.10	757	0.04	472	0.13	203	0.08
5	3129	0.07	2060	0.07	761	0.01	440	0.08	214	0.14
6	3181	0.06	2097	0.06	685	0.01	438	0.04	294	−0.14
7	3129	0.07	1996	0.08	721	0.01	470	0.02	208	0.22
8	3171	0.04	2085	−0.00	707	−0.08	454	−0.10	248	−0.12
9	3097	0.08	2016	0.04	675	0.07	480	−0.02	202	0.16
High	2945	0.15	1938	0.07	654	0.07	398	−0.12	236	−0.15

continued

Table 4.4 Continued

Portfolio	MAP(5)		MAP(10)		MAP(60)		MAP(120)		MAP(250)	
	NT	BETC	NT	BETC	NT	BETC	NT	BETC	NT	BETC
				Panel D: Short-term reversal sorted portfolios						
Low	2926	−0.04	1890	−0.14	607	−0.75	328	−1.17	200	−1.70
2	2989	0.08	2020	0.02	701	−0.26	426	−0.31	254	−0.63
3	3015	0.10	2029	0.07	705	−0.04	472	−0.18	318	−0.47
4	3041	0.11	1937	0.13	673	0.06	454	−0.01	300	−0.25
5	3139	0.07	2041	0.07	713	0.06	480	0.04	256	−0.12
6	3149	0.07	2044	0.05	663	0.09	424	0.02	228	−0.14
7	3189	0.05	2101	0.06	727	0.12	482	0.12	236	0.40
8	3187	0.08	2049	0.10	693	0.22	483	0.27	266	0.35
9	3232	0.09	2052	0.09	809	0.14	519	0.28	276	0.38
High	3311	0.24	2179	0.31	783	0.85	501	1.39	373	1.55
				Panel E: Long-term reversal sorted portfolios						
Low	3103	0.15	1908	0.23	713	0.16	458	0.14	280	−0.24
2	3103	0.09	1934	0.11	725	0.05	402	−0.00	276	−0.26
3	3113	0.07	2062	0.05	677	−0.01	450	−0.06	271	−0.28
4	3109	0.05	2077	−0.00	683	−0.03	482	−0.09	212	−0.07
5	3157	0.05	2049	0.05	663	0.03	482	−0.07	220	−0.02
6	3139	0.04	2018	0.02	697	−0.03	448	−0.11	260	−0.05
7	3199	0.05	2089	0.04	713	−0.01	460	−0.05	236	−0.10
8	3179	0.06	2062	0.05	709	0.01	502	0.02	286	−0.11
9	3111	0.11	2027	0.10	755	0.09	492	0.14	262	0.21
High	3079	0.16	1996	0.16	725	0.25	459	0.28	272	0.23
				Panel F: Volatility-sorted portfolios						
High	2580	0.33	1558	0.52	357	1.14	260	0.58	124	0.24
2	2653	0.40	1650	0.53	505	0.81	330	0.64	165	0.57
3	2717	0.34	1634	0.48	509	0.73	314	0.77	190	0.65
4	2695	0.27	1674	0.36	515	0.57	360	0.42	163	0.43
5	2636	0.26	1624	0.34	539	0.42	336	0.38	138	0.46
6	2639	0.25	1646	0.32	541	0.35	359	0.28	198	0.12
7	2613	0.22	1664	0.27	507	0.37	300	0.34	158	0.27
8	2661	0.20	1640	0.25	447	0.42	334	0.29	144	0.27
9	2569	0.20	1546	0.27	418	0.46	270	0.40	174	0.17
Low	2252	0.23	1350	0.32	360	0.51	228	0.50	105	0.73

continued

Table 4.4 Continued

Portfolio	MAP(5)		MAP(10)		MAP(60)		MAP(120)		MAP(250)	
	NT	BETC	NT	BETC	NT	BETC	NT	BETC	NT	BETC
					Panel G: Industry-sorted portfolios					
NoDur	3043	0.08	1978	0.09	662	0.08	442	0.07	238	−0.00
Durbl	3235	0.13	2083	0.13	751	0.23	470	0.27	290	0.13
Manuf	3013	0.13	1935	0.14	667	0.17	458	0.09	218	−0.01
Enrgy	3197	0.04	2125	−0.02	801	−0.19	599	−0.28	306	−0.49
HiTec	3229	0.08	2123	0.10	781	0.07	537	0.09	318	0.03
Telcm	3386	0.02	2218	−0.00	793	−0.03	510	0.00	276	0.03
Shops	3078	0.13	1942	0.17	664	0.14	449	0.04	341	−0.18
Hlth	3153	0.09	2048	0.09	717	−0.02	497	−0.06	319	−0.20
Utils	2941	0.14	1838	0.18	572	0.19	402	0.10	194	0.28
Other	2901	0.19	1930	0.19	623	0.25	454	0.17	218	0.19

MA switching strategy is quite successful at improving the returns over a buy-and-hold investment strategy. The superior performance is robust with respect to two subperiods, various lag lengths of the MA window, and persists for between 5 and 250 days with very reasonable intensity of trading and substantial BETC. This suggests that the MA switching strategy will be of use to large institutional investors. These findings are perhaps suggestive as to the wide popularity of the MA as a technical indicator in practice.

4.2 Short-Selling

In this case, the returns of the MA switching strategy can be expressed as follows:

$$\tilde{R}_{jt,L} = \begin{cases} R_{jt}, & \text{if } P_{jt-1} > A_{jt-1,L}, \\ -R_{jt}, & \text{otherwise,} \end{cases} \tag{4.1}$$

in the absence of any transaction costs imposed on the switches. The alternative specification for the case of positive one-way transaction cost of τ leads to the following four cases in the post-transaction cost returns:

$$\tilde{R}_{jt,L} = \begin{cases} R_{jt}, & \text{if } P_{jt-1} > A_{jt-1,L} \text{ and } P_{jt-2} > A_{jt-2,L}, \\ R_{jt} - \tau, & \text{if } P_{jt-1} > A_{jt-1,L} \text{ and } P_{jt-2} < A_{jt-2,L}, \\ -R_{jt}, & \text{if } P_{jt-1} < A_{jt-1,L} \text{ and } P_{jt-2} < A_{jt-2,L}, \\ -R_{jt} - \tau, & \text{if } P_{jt-1} < A_{jt-1,L} \text{ and } P_{jt-2} > A_{jt-2,L}. \end{cases} \tag{4.2}$$

Table 4.5 presents the summary statistics for the performance of the MA, BH, and MAP returns when short-selling is permitted. It is instructive to compare these to the results for the baseline implementation reported in Table 2.1. As expected, permitting short-selling on a negative signal magnifies both the positive returns and the negative returns. In general, the positive returns in the baseline implementation are magnified by a factor of almost 2 when short-selling is permitted. However, the negative returns in the baseline case are magnified by a larger factor. This case sometimes leads to the underperformance of the MA strategy relative to the BH strategy as evidenced by the negative average MAP returns in Table 4.5 for some portfolios.

Table 4.6 reports the parameter estimates for the CAPM model, the Fama-French three-factor, model, and the Carhart four-factor model. Once again, there is overwhelming presence of positive and statistically significant abnormal returns. The level of annualized αs can be as high as 34% for the smallest capitalization decile when using the four-factor model. This leads to the conclusion that short-selling, though riskier than just holding the risk-free asset, still generates substantial risk-adjusted returns compared to the buy-and-hold strategy.

Table 4.7 compiles the findings regarding the relative performance of the MA strategy versus the BH strategy as well as the trading intensity involved. Overall, the picture that emerges is that when the strategy works well, short-selling can really improve performance to as much as 27.5% for the loser momentum portfolio. However, when the strategy does not work so well, short-selling only serves to magnify losses, in this case to as low as −14.4% for the smallest short-term reversal portfolio.

Finally, Table 4.8 presents the regression estimates for the two most commonly used market timing regressions. The coefficients of primary interest here are β_{m^2} and γ_m. In the vast majority of the cases the estimates for these coefficients are positive and statistically significant, indicating that the MA strategy has substantial market timing power. However, a few portfolios do have negative albeit

Table 4.5 Portfolio performance with short-selling

This table reports summary statistics for the respective buy and hold (BH) portfolio returns, the MA switching strategy portfolio returns where short-selling is permitted conditional on a sell signal, and the excess return of MA over BH (MAP) using sets of ten portfolios sorted by various characteristics. The sample period covers January 4, 1960 until December 12, 2013 with value-weighted portfolio returns. μ the annualized average return, σ is annualized standard deviation of returns, s is the annualized skewness, k is the annualized kurtosis, and SR is the annualized Sharpe ratio. The length of the moving average window is 20 days.

Portfolio	BH portfolios					MA portfolios					MAP portfolios				
	μ	σ	s	k	SR	μ	σ	s	k	SR	μ	σ	s	k	SR
Panel A: Size-sorted portfolios															
Low	11.88	13.58	−0.90	14.81	0.52	35.69	13.42	0.41	14.93	2.30	23.81	21.20	1.49	34.15	1.12
2	11.86	16.72	−0.46	13.85	0.52	28.23	16.64	−0.10	14.02	2.30	16.37	25.85	0.47	28.76	1.12
3	12.96	16.60	−0.49	12.02	0.42	26.40	16.53	0.16	12.00	1.41	13.44	25.67	0.76	27.37	0.63
4	12.18	16.33	−0.48	12.48	0.49	24.27	16.28	0.01	12.51	1.31	12.09	25.23	0.59	27.61	0.52
5	12.68	16.23	−0.47	11.96	0.45	21.77	16.19	−0.05	11.97	1.20	9.09	24.99	0.51	26.43	0.48
6	12.26	15.33	−0.53	13.43	0.49	21.06	15.29	0.05	13.41	1.05	8.80	23.62	0.68	31.11	0.36
7	12.29	15.47	−0.55	15.54	0.49	20.45	15.44	0.10	15.51	1.07	8.16	23.87	0.76	36.49	0.37
8	12.01	15.70	−0.51	16.62	0.49	17.31	15.68	0.11	16.56	1.02	5.31	24.18	0.70	40.02	0.34
9	11.57	15.46	−0.59	20.90	0.46	12.01	15.46	0.34	20.73	0.80	0.45	24.00	1.00	52.11	0.22
High	10.23	16.00	−0.50	21.64	0.44	3.91	16.01	0.32	21.49	0.47	−6.32	24.75	0.84	54.11	0.02

Panel B: Book-to-market sorted portfolios

Low	9.49	17.53	−0.18	13.47	0.27	11.25	17.52	0.03	13.45	0.37	1.76	27.39	0.23	32.00	0.06
2	11.03	16.07	−0.39	16.37	0.27	10.95	16.07	0.16	16.27	0.37	−0.09	24.76	0.60	40.34	0.06
3	11.15	15.42	−0.52	20.70	0.39	11.72	15.42	0.33	20.55	0.38	0.57	23.59	0.95	54.48	−0.00
4	10.95	15.80	−0.71	20.94	0.41	12.26	15.80	0.44	20.75	0.45	1.31	24.10	1.31	53.09	0.02
5	11.13	15.62	−0.45	23.02	0.39	7.86	15.62	0.08	22.88	0.47	−3.27	23.83	0.58	57.72	0.05
6	11.89	15.14	−0.45	16.96	0.41	11.16	15.14	0.12	16.84	0.20	−0.73	23.05	0.63	42.17	−0.14
7	12.55	14.89	−0.64	22.73	0.47	11.60	14.90	0.29	22.52	0.42	−0.95	22.51	1.07	61.04	−0.03
8	12.98	15.82	−0.60	30.16	0.52	11.57	15.82	0.18	29.97	0.46	−1.41	24.06	0.88	72.10	−0.04
9	13.90	15.99	−0.68	20.87	0.52	13.48	15.99	0.22	20.66	0.43	−0.43	24.14	1.04	53.74	−0.06
High	15.01	17.78	−0.46	15.88	0.57	18.51	17.76	0.09	15.79	0.54	3.51	26.68	0.66	39.37	−0.02

continued

Table 4.5 Continued

Portfolio	BH portfolios					MA portfolios					MAP portfolios				
	μ	σ	s	k	SR	μ	σ	s	k	SR	μ	σ	s	k	SR
Panel C: Momentum-sorted portfolios															
Low	2.59	24.93	0.41	25.70	−0.09	30.15	24.86	0.02	25.97	1.02	27.55	38.38	−0.34	45.73	0.72
2	8.03	20.35	0.15	22.17	−0.09	20.81	20.31	−0.32	22.41	1.02	12.78	30.80	−0.49	45.30	0.72
3	10.29	17.59	0.04	17.98	0.16	17.29	17.57	0.01	18.06	0.79	7.00	26.25	0.00	38.06	0.41
4	10.22	16.61	−0.15	18.81	0.31	13.42	16.60	0.25	18.77	0.71	3.20	25.13	0.48	46.26	0.27
5	10.09	15.89	0.01	21.65	0.33	9.76	15.90	−0.02	21.65	0.52	−0.34	23.89	−0.04	52.44	0.13
6	10.99	15.35	−0.58	26.71	0.33	9.77	15.35	0.26	26.54	0.31	−1.22	23.25	0.96	70.31	−0.01
7	10.92	15.19	−0.55	22.88	0.40	9.07	15.20	0.29	22.71	0.32	−1.85	23.24	0.93	60.32	−0.05
8	13.57	15.49	−0.73	21.21	0.40	9.31	15.51	0.38	20.93	0.28	−4.27	23.47	1.26	57.86	−0.08
9	12.81	16.47	−0.54	15.10	0.57	10.28	16.48	0.21	14.94	0.29	−2.52	25.52	0.79	37.23	−0.18
High	17.85	20.37	−0.50	12.74	0.49	18.17	20.37	0.03	12.62	0.33	0.32	31.93	0.55	29.81	−0.10

Panel D: Short-term reversal sorted portfolios

Low	41.02	25.55	0.87	29.76	1.42	26.59	25.63	−1.13	30.09	0.85	−14.43	41.28	−1.95	64.00	−0.35
2	18.83	20.36	0.33	27.47	1.42	15.44	20.38	−0.50	27.59	0.85	−3.38	33.36	−0.77	55.57	−0.35
3	14.32	17.97	0.06	26.89	0.69	12.69	17.98	−0.37	26.94	0.52	−1.64	29.07	−0.41	56.28	−0.10
4	11.57	16.65	−0.38	23.86	0.53	13.47	16.65	0.21	23.78	0.44	1.90	26.48	0.60	53.07	−0.06
5	11.73	15.81	−0.47	22.77	0.41	12.34	15.81	0.32	22.63	0.52	0.61	24.93	0.81	53.78	0.07
6	12.78	15.57	−0.45	22.73	0.44	13.01	15.57	0.26	22.59	0.48	0.23	23.72	0.80	60.78	0.02
7	9.63	15.45	−0.55	19.99	0.51	9.51	15.45	0.27	19.86	0.53	−0.11	23.16	0.97	53.93	0.01
8	7.93	15.66	−0.42	16.68	0.31	10.00	15.65	0.17	16.62	0.31	2.07	22.68	0.79	44.32	−0.00
9	6.01	16.52	−0.44	15.75	0.20	6.63	16.52	−0.12	15.73	0.33	0.62	23.32	0.46	39.80	0.09
High	−9.59	19.08	−0.55	11.88	0.07	6.76	19.09	−0.08	11.95	0.11	16.34	28.31	0.66	22.13	0.03

continued

Table 4.5 Continued

Portfolio	BH portfolios					MA portfolios					MAP portfolios				
	μ	σ	s	k	SR	μ	σ	s	k	SR	μ	σ	s	k	SR
Panel E: Long-term reversal sorted portfolios															
Low	14.73	20.02	−0.38	14.61	0.50	23.82	19.98	0.31	14.53	0.95	9.09	30.30	0.83	37.12	0.30
2	13.72	17.10	−0.61	18.75	0.50	14.37	17.10	0.25	18.59	0.95	0.65	25.94	0.99	48.47	0.30
3	13.09	15.94	−0.58	18.32	0.52	11.22	15.95	0.33	18.12	0.56	−1.87	23.78	1.08	51.10	0.03
4	12.96	15.66	−0.79	25.23	0.52	9.32	15.67	0.52	24.92	0.40	−3.64	23.49	1.53	70.99	−0.08
5	11.63	15.50	−0.71	22.89	0.52	10.08	15.50	0.51	22.65	0.29	−1.55	23.24	1.44	61.32	−0.15
6	12.91	14.73	−0.36	21.45	0.44	9.90	14.74	0.10	21.30	0.34	−3.02	22.09	0.53	60.02	−0.07
7	12.10	15.03	−0.39	17.81	0.55	9.46	15.04	0.16	17.67	0.35	−2.64	22.51	0.64	48.54	−0.14
8	11.03	15.60	−0.46	19.22	0.49	9.54	15.60	0.27	19.08	0.31	−1.50	24.04	0.80	48.15	−0.12
9	9.75	16.87	−0.19	18.39	0.40	10.40	16.87	−0.03	18.37	0.30	0.65	26.28	0.17	44.08	−0.06
High	10.66	20.33	−0.16	14.88	0.29	14.39	20.33	−0.12	14.90	0.33	3.73	32.20	0.06	32.89	0.02

Panel F: Volatility-sorted portfolios

High	40.10	20.27	0.10	14.40	1.74	62.29	20.04	0.33	14.71	2.87	22.18	28.83	0.44	38.43	0.77
2	17.80	19.30	−0.31	14.52	1.74	41.05	19.16	0.28	14.64	2.87	23.25	29.63	0.75	33.20	0.77
3	15.22	18.25	−0.38	16.39	0.67	35.47	18.14	0.11	16.60	1.89	20.25	28.19	0.61	36.62	0.78
4	15.74	16.89	−0.47	18.77	0.57	30.35	16.81	0.10	18.91	1.69	14.61	26.00	0.70	42.16	0.72
5	15.68	15.52	−0.55	20.47	0.65	28.59	15.44	0.09	20.62	1.52	12.91	23.84	0.78	46.43	0.56
6	14.82	14.38	−0.67	24.45	0.70	24.56	14.33	0.08	24.56	1.54	9.74	22.10	0.89	56.32	0.54
7	14.40	12.78	−0.66	28.32	0.70	23.59	12.72	0.15	28.50	1.38	9.20	19.72	0.93	70.12	0.44
8	13.68	11.15	−0.78	36.32	0.75	23.30	11.09	0.46	36.60	1.48	9.62	17.39	1.36	92.78	0.47
9	12.53	9.27	−0.58	54.72	0.80	21.97	9.20	0.29	55.95	1.67	9.45	14.63	0.96	136.30	0.55
Low	10.81	6.86	−0.47	71.59	0.84	19.73	6.78	0.27	74.49	1.87	8.92	10.85	0.82	177.24	0.65

continued

Table 4.5 Continued

Portfolio	BH portfolios					MA portfolios					MAP portfolios				
	μ	σ	s	k	SR	μ	σ	s	k	SR	μ	σ	s	k	SR
Panel G: Industry-sorted portfolios															
NoDur	12.80	13.63	−0.62	22.17	0.59	14.85	13.62	0.34	21.97	0.74	2.05	20.37	1.17	63.09	0.10
Durbl	10.57	20.63	−0.23	11.77	0.59	11.47	20.63	0.21	11.71	0.74	0.90	30.76	0.54	28.39	0.10
Manuf	11.07	16.28	−0.68	21.25	0.28	14.68	16.27	0.38	21.10	0.32	3.60	24.91	1.20	54.42	0.03
Enrgy	13.49	19.99	−0.20	19.86	0.39	7.62	20.00	−0.06	19.79	0.61	−5.87	29.59	0.15	56.37	0.14
HiTec	11.37	22.42	0.02	12.00	0.44	15.32	22.41	−0.06	12.03	0.14	3.94	34.77	−0.08	28.50	−0.20
Telcm	10.94	17.36	−0.03	17.12	0.29	7.01	17.37	−0.00	17.09	0.47	−3.93	26.03	0.02	45.75	0.11
Shops	12.23	16.54	−0.28	14.21	0.36	15.60	16.53	0.14	14.15	0.13	3.37	24.96	0.50	36.48	−0.15
Hlth	12.52	16.83	−0.40	14.37	0.45	12.71	16.83	0.19	14.26	0.65	0.19	25.29	0.70	38.21	0.13
Utils	10.18	12.93	−0.03	28.88	0.46	14.03	12.91	−0.29	29.07	0.47	3.85	19.74	−0.27	76.72	0.01
Other	11.15	17.73	−0.26	18.07	0.42	16.23	17.72	0.03	18.08	0.72	5.09	27.49	0.34	37.22	0.20

Table 4.6 Factor loadings with short-selling

This table reports alphas, betas, and adjusted R^2 of the regressions of the MAP excess returns on the market factor, the Fama-French three-factors, and the Carhart four-factors using portfolios sorted by various characteristics. The alphas are annualized and in percent. The sample period covers January 4, 1960 until December 31, 2013 with daily value-weighted portfolio returns. The length of the MA window is 20 days. Newey and West (1987) standard errors with five lags are used in reporting statistical significance of a two-sided null hypothesis at the 1%, 5%, and 10% level given by a***, a**, and a*, respectively.

	CAPM			Fama-French					Carhart					
Portfolio	α	β_m	$\overline{R}^2$	α	β_m	β_s	β_h	$\overline{R}^2$	α	β_m	β_s	β_h	β_u	$\overline{R}^2$
Panel A: Size-sorted portfolios														
Low	29.063***	−0.870***	0.405	32.391***	−0.983***	−1.045***	−0.266***	0.554	34.244***	−1.007***	−1.040***	−0.328***	−0.173***	0.561
2	23.144***	−1.122***	0.453	26.930***	−1.250***	−1.199***	−0.299***	0.585	28.842***	−1.275***	−1.194***	−0.363***	−0.179***	0.590
3	20.382***	−1.149***	0.483	23.771***	−1.265***	−1.111***	−0.255***	0.596	25.530***	−1.288***	−1.106***	−0.313***	−0.164***	0.601
4	18.993***	−1.143***	0.494	21.727***	−1.239***	−1.001***	−0.170***	0.589	23.736***	−1.265***	−0.996***	−0.237***	−0.188***	0.595
5	16.037***	−1.151***	0.511	18.210***	−1.230***	−0.898***	−0.100***	0.588	19.825***	−1.251***	−0.894***	−0.154***	−0.151***	0.592
6	15.481***	−1.107***	0.528	16.854***	−1.158***	−0.630***	−0.042**	0.571	18.575***	−1.180***	−0.626***	−0.099***	−0.161***	0.576
7	14.986***	−1.130***	0.540	16.352***	−1.179***	−0.513***	−0.080***	0.568	18.184***	−1.202***	−0.508***	−0.142***	−0.171***	0.573
8	12.341***	−1.165***	0.559	13.559***	−1.206***	−0.359***	−0.105***	0.573	15.289***	−1.228***	−0.354***	−0.163***	−0.162***	0.578
9	7.574***	−1.180***	0.583	8.488***	−1.207***	−0.123***	−0.129***	0.585	10.189***	−1.229***	−0.118***	−0.186***	−0.159***	0.590
High	1.073	−1.224***	0.588	−0.106	−1.185***	0.294***	0.120***	0.597	0.919	−1.199***	0.297***	0.086***	−0.096***	0.599

continued

Table 4.6 Continued

Portfolio	CAPM			Fama-French					Carhart					
	α	β_m	$\bar{R}^2$	α	β_m	β_s	β_h	$\bar{R}^2$	α	β_m	β_s	β_h	β_u	$\bar{R}^2$
Panel B: Book-to-market sorted portfolios														
Low	9.764***	−1.325***	0.563	6.189***	−1.229***	0.081***	0.640***	0.591	7.086***	−1.240***	0.083***	0.610***	−0.084***	0.592
2	7.149***	−1.199***	0.564	5.944***	−1.166***	0.050***	0.208***	0.568	7.477***	−1.185***	0.054***	0.157***	−0.143***	0.572
3	7.350***	−1.123***	0.546	6.880***	−1.109***	0.048***	0.071***	0.546	9.009***	−1.137***	0.054***	0.000	−0.199***	0.554
4	8.238***	−1.147***	0.546	9.636***	−1.184***	−0.010	−0.258***	0.552	11.106***	−1.203***	−0.006	−0.307***	−0.137***	0.555
5	3.414	−1.107***	0.519	5.351**	−1.158***	−0.019	−0.355***	0.531	6.556***	−1.174***	−0.015	−0.396***	−0.113***	0.533
6	5.723**	−1.068***	0.517	7.881***	−1.129***	−0.140***	−0.355***	0.531	9.082***	−1.144***	−0.137***	−0.395***	−0.112***	0.533
7	5.244**	−1.025***	0.500	8.033***	−1.101***	−0.063***	−0.499***	0.525	9.776***	−1.123***	−0.058***	−0.558***	−0.163***	0.531
8	4.996**	−1.062***	0.469	9.859***	−1.193***	−0.144***	−0.859***	0.534	10.592***	−1.203***	−0.142***	−0.883***	−0.068***	0.535
9	6.093**	−1.080***	0.482	10.331***	−1.195***	−0.127***	−0.748***	0.531	12.386***	−1.221***	−0.122***	−0.817***	−0.192***	0.538
High	10.528***	−1.163***	0.457	15.795***	−1.311***	−0.369***	−0.858***	0.517	17.737***	−1.336***	−0.364***	−0.923***	−0.181***	0.522

						Panel C: Momentum-sorted portfolios								
Low	37.581***	−1.661***	0.451	40.697***	−1.756***	−0.548***	−0.395***	0.467	31.402***	−1.636***	−0.572***	−0.084**	0.868***	0.522
2	21.194***	−1.394***	0.493	22.951***	−1.445***	−0.162***	−0.273***	0.498	16.361***	−1.359***	−0.179***	−0.053**	0.615***	0.542
3	14.296***	−1.208***	0.510	15.699***	−1.246***	−0.003	−0.261***	0.515	11.668***	−1.193***	−0.013	−0.126***	0.376***	0.537
4	10.340***	−1.182***	0.533	11.560***	−1.213***	0.041**	−0.242***	0.538	10.050***	−1.194***	0.037**	−0.191***	0.141***	0.541
5	6.415***	−1.118***	0.527	7.188***	−1.139***	0.001	−0.145***	0.529	7.171***	−1.138***	0.001	−0.144***	0.002	0.529
6	5.303**	−1.081***	0.521	5.855***	−1.094***	0.052***	−0.121***	0.522	7.184***	−1.111***	0.056***	−0.165***	−0.124***	0.525
7	4.742**	−1.092***	0.532	5.301**	−1.106***	0.024	−0.112***	0.533	8.124***	−1.143***	0.031*	−0.207***	−0.264***	0.547
8	2.267	−1.082***	0.512	2.906	−1.098***	0.037**	−0.132***	0.514	7.285***	−1.155***	0.048***	−0.278***	−0.409***	0.546
9	4.669*	−1.191***	0.524	5.071**	−1.206***	−0.156***	−0.022	0.527	10.678***	−1.278***	−0.141***	−0.210***	−0.524***	0.572
High	8.866***	−1.416***	0.473	7.847**	−1.400***	−0.467***	0.349***	0.494	16.615***	−1.513***	−0.444***	0.056**	−0.819***	0.565

continued

Table 4.6 Continued

	CAPM			Fama-French					Carhart					
Portfolio	α	β_m	$\overline{R}^2$	α	β_m	β_s	β_h	$\overline{R}^2$	α	β_m	β_s	β_h	β_u	$\overline{R}^2$
							Panel D: Short-term reversal sorted portfolios							
Low	−4.386	−1.663***	0.391	−3.343	−1.702***	−0.466***	−0.036	0.398	−5.127	−1.679***	−0.470***	0.024	0.167***	0.400
2	5.746*	−1.512***	0.494	6.390**	−1.533***	−0.189***	−0.056**	0.496	6.450**	−1.534***	−0.188***	−0.058**	−0.006	0.496
3	6.433**	−1.337***	0.509	7.010**	−1.354***	−0.099***	−0.074***	0.510	8.348***	−1.372***	−0.095***	−0.119***	−0.125***	0.512
4	9.423***	−1.246***	0.533	9.813***	−1.257***	−0.015	−0.068***	0.534	11.531***	−1.279***	−0.010	−0.125***	−0.160***	0.538
5	7.810***	−1.192***	0.551	8.340***	−1.207***	−0.022	−0.091***	0.552	9.847***	−1.226***	−0.018	−0.142***	−0.141***	0.555
6	7.002***	−1.122***	0.538	7.553***	−1.136***	−0.009	−0.100***	0.539	9.188***	−1.157***	−0.004	−0.155***	−0.153***	0.544
7	6.612***	−1.114***	0.557	7.208***	−1.130***	−0.022	−0.104***	0.558	8.578***	−1.148***	−0.019	−0.150***	−0.128***	0.561
8	8.593***	−1.080***	0.546	9.003***	−1.092***	−0.034*	−0.065***	0.546	10.990***	−1.117***	−0.028*	−0.132***	−0.186***	0.554
9	7.222***	−1.093***	0.529	8.068***	−1.117***	−0.090***	−0.127***	0.531	9.069***	−1.130***	−0.087***	−0.161***	−0.093***	0.533
High	24.320***	−1.321***	0.524	25.269***	−1.355***	−0.365***	−0.053**	0.534	25.319***	−1.356***	−0.365***	−0.055**	−0.005	0.534

Panel E: Long-term reversal sorted portfolios

Low	17.286***	−1.357***	0.483	20.757***	−1.465***	−0.671***	−0.420***	0.519	23.748***	−1.503***	−0.663***	−0.520***	−0.279***	0.528
2	7.855***	−1.193***	0.509	10.288***	−1.264***	−0.298***	−0.353***	0.525	13.116***	−1.301***	−0.291***	−0.447***	−0.264***	0.536
3	4.697**	−1.088***	0.503	6.741***	−1.144***	−0.118***	−0.341***	0.515	9.194***	−1.176***	−0.112***	−0.423***	−0.229***	0.525
4	2.843	−1.074***	0.503	5.068**	−1.134***	−0.055***	−0.397***	0.517	6.630***	−1.154***	−0.051***	−0.449***	−0.146***	0.521
5	4.890**	−1.066***	0.507	7.032***	−1.123***	−0.008	−0.397***	0.522	8.605***	−1.143***	−0.004	−0.450***	−0.147***	0.526
6	3.119	−1.016***	0.510	4.057*	−1.039***	0.097***	−0.208***	0.516	5.908***	−1.063***	0.102***	−0.270***	−0.173***	0.522
7	3.647*	−1.041***	0.515	4.088*	−1.051***	0.089***	−0.113***	0.518	5.815***	−1.073***	0.094***	−0.170***	−0.161***	0.523
8	5.369**	−1.137***	0.539	5.286**	−1.132***	0.116***	−0.024	0.541	7.452***	−1.161***	0.121***	−0.096***	−0.202***	0.548
9	8.144***	−1.242***	0.537	7.577***	−1.224***	0.110***	0.069***	0.538	9.380***	−1.247***	0.114***	0.008	−0.168***	0.543
High	12.964***	−1.529***	0.543	10.357***	−1.463***	−0.133***	0.532***	0.559	10.872***	−1.470***	−0.132***	0.515***	−0.048***	0.559

continued

Table 4.6 Continued

	CAPM			Fama-French					Carhart					
Portfolio	α	β_m	$\bar{R}^2$	α	β_m	β_s	β_h	$\bar{R}^2$	α	β_m	β_s	β_h	β_u	$\bar{R}^2$
							Panel F: Volatility-sorted portfolios							
High	28.847***	−1.104***	0.353	32.368***	−1.222***	−1.064***	−0.295***	0.437	33.683***	−1.239***	−1.060***	−0.339***	−0.123***	0.439
2	31.085***	−1.298***	0.462	34.489***	−1.412***	−1.004***	−0.294***	0.533	35.525***	−1.425***	−1.001***	−0.329***	−0.097***	0.534
3	27.852***	−1.259***	0.480	31.297***	−1.372***	−0.902***	−0.336***	0.546	32.718***	−1.390***	−0.898***	−0.384***	−0.133***	0.548
4	21.659***	−1.167***	0.485	25.072***	−1.275***	−0.757***	−0.379***	0.543	26.267***	−1.291***	−0.754***	−0.419***	−0.112***	0.545
5	19.373***	−1.070***	0.485	22.656***	−1.171***	−0.613***	−0.404***	0.535	23.864***	−1.187***	−0.609***	−0.445***	−0.113***	0.537
6	15.758***	−0.997***	0.490	18.730***	−1.087***	−0.476***	−0.393***	0.530	19.882***	−1.101***	−0.473***	−0.431***	−0.108***	0.533
7	14.538***	−0.885***	0.485	16.752***	−0.951***	−0.328***	−0.302***	0.511	18.109***	−0.969***	−0.324***	−0.347***	−0.127***	0.515
8	14.256***	−0.768***	0.469	16.055***	−0.820***	−0.221***	−0.261***	0.488	17.410***	−0.838***	−0.218***	−0.306***	−0.127***	0.494
9	13.111***	−0.607***	0.414	14.643***	−0.651***	−0.142***	−0.238***	0.432	15.719***	−0.665***	−0.139***	−0.274***	−0.101***	0.437
Low	11.002***	−0.345***	0.244	11.553***	−0.361***	−0.042***	−0.089***	0.248	12.580***	−0.374***	−0.039***	−0.123***	−0.096***	0.256

Panel G: Industry-sorted portfolios

NoDur	7.277***	−0.866***	0.435	7.042***	−0.857***	0.133***	−0.001	0.438	9.354***	−0.887***	0.139***	−0.078***	−0.216***	0.450
Durbl	9.060***	−1.351***	0.464	11.318***	−1.412***	−0.074***	−0.396***	0.473	12.359***	−1.426***	−0.071***	−0.431***	−0.097***	0.474
Manuf	10.746***	−1.183***	0.543	11.535***	−1.205***	−0.039**	−0.134***	0.545	13.648***	−1.232***	−0.033*	−0.205***	−0.197***	0.551
Enrgy	0.824	−1.108***	0.338	2.082	−1.136***	0.239***	−0.316***	0.349	5.085	−1.175***	0.247***	−0.417***	−0.280***	0.358
HiTec	13.417***	−1.569***	0.490	8.506***	−1.441***	−0.086***	0.946***	0.529	8.308***	−1.439***	−0.087***	0.953***	0.019	0.529
Telcm	2.574	−1.078***	0.413	3.355	−1.093***	0.226***	−0.223***	0.422	3.547	−1.095***	0.227***	−0.229***	−0.018	0.422
Shops	10.011***	−1.100***	0.467	9.667***	−1.092***	−0.061***	0.085***	0.468	11.243***	−1.113***	−0.057***	0.032	−0.147***	0.472
Hlth	6.524**	−1.048***	0.414	4.765*	−0.998***	0.186***	0.265***	0.422	6.993***	−1.026***	0.192***	0.191***	−0.208***	0.429
Utils	8.057***	−0.696***	0.299	10.222***	−0.750***	0.134***	−0.450***	0.331	11.928***	−0.772***	0.139***	−0.507***	−0.159***	0.338
Other	12.918***	−1.297***	0.536	16.042***	−1.386***	−0.270***	−0.491***	0.557	16.370***	−1.390***	−0.269***	−0.502***	−0.031*	0.557

Table 4.7 Trading intensity with short-selling
This table reports the results for the improvement delivered by the MA switching strategy over the buy-and-hold strategy when short-selling is permitted, the trading frequency, as well as the BETC using ten decile portfolios sorted by various characteristics. The sample period covers January 4, 1960 until December 31, 2013 with value-weighted portfolio returns. $\Delta\mu$ is the annualized improvement in the average in-sample monthly return, $\Delta\sigma$ is the annualized improvement in the return standard deviation, p_A is the proportion of months during which there is a hold signal, NT is the number of transactions (buy or sell) over the entire sample period, BETC is the break-even one-sided transaction cost in percent, p_1 is the proportion of months during which a buy signal was followed by a positive return of the underlying portfolio, and p_2 is the proportion of months during which a buy signal was followed by a portfolio return in excess of the risk-free rate. The length of the MA window is 20 days.

Portfolio	$\Delta\mu$	$\Delta\sigma$	p_A	NT	BETC	p_1	p_2
			Panel A: Size-sorted portfolios				
Low	23.81	0.17	0.60	931	1.39	60.19	57.78
2	16.37	0.08	0.61	1071	0.83	57.70	56.44
3	13.44	0.06	0.61	1089	0.67	56.60	56.48
4	12.09	0.05	0.62	1147	0.57	56.52	56.08
5	9.09	0.04	0.62	1139	0.43	55.71	56.05
6	8.80	0.04	0.62	1165	0.41	55.87	55.84
7	8.16	0.03	0.62	1145	0.39	55.93	55.50
8	5.31	0.02	0.62	1193	0.24	54.52	54.99
9	0.45	0.00	0.62	1313	0.02	53.47	54.43
High	−6.32	−0.01	0.61	1501	−0.23	50.46	52.85
			Panel B: Book-to-market sorted portfolios				
Low	1.76	0.00	0.59	1409	0.07	52.24	52.84
2	−0.09	−0.00	0.61	1339	−0.00	52.41	53.19
3	0.57	0.00	0.61	1345	0.02	52.44	53.07
4	1.31	0.00	0.61	1315	0.05	53.02	53.69
5	−3.27	−0.01	0.61	1407	−0.13	51.54	53.15
6	−0.73	−0.00	0.62	1259	−0.03	52.95	53.77
7	−0.95	−0.00	0.62	1301	−0.04	53.21	53.95
8	−1.41	−0.00	0.63	1315	−0.06	52.79	54.23
9	−0.43	−0.00	0.63	1309	−0.02	53.21	54.11
High	3.51	0.01	0.62	1281	0.15	53.64	53.97

continued

Table 4.7 Continued

Portfolio	$\Delta\mu$	$\Delta\sigma$	p_A	NT	BETC	p_1	p_2
Panel C: Momentum-sorted portfolios							
Low	27.55	0.07	0.52	1197	1.25	54.87	50.78
2	12.78	0.04	0.56	1295	0.54	53.53	51.25
3	7.00	0.02	0.58	1315	0.29	52.78	52.40
4	3.20	0.01	0.59	1359	0.13	52.18	52.55
5	−0.34	−0.00	0.60	1361	−0.01	52.25	52.98
6	−1.22	−0.00	0.62	1401	−0.05	52.65	53.48
7	−1.85	−0.00	0.62	1355	−0.07	52.29	53.58
8	−4.27	−0.01	0.63	1389	−0.17	52.44	54.65
9	−2.52	−0.01	0.62	1435	−0.10	52.58	54.20
High	0.32	0.00	0.64	1307	0.01	54.37	55.74
Panel D: Short-term reversal sorted portfolios							
Low	−14.43	−0.08	0.72	1241	−0.63	56.74	57.69
2	−3.38	−0.01	0.64	1277	−0.14	54.39	54.68
3	−1.64	−0.00	0.62	1345	−0.07	53.58	53.64
4	1.90	0.01	0.61	1299	0.08	53.19	53.37
5	0.61	0.00	0.61	1355	0.02	52.66	53.45
6	0.23	0.00	0.62	1315	0.01	52.84	53.56
7	−0.11	−0.00	0.60	1413	−0.00	52.10	53.13
8	2.07	0.00	0.58	1401	0.08	51.51	52.97
9	0.62	0.00	0.57	1409	0.02	51.72	53.08
High	16.34	−0.00	0.47	1475	0.60	50.75	50.48
Panel E: Long-term reversal sorted portfolios							
Low	9.09	0.04	0.58	1295	0.38	53.61	53.33
2	0.65	0.00	0.61	1309	0.03	53.09	53.92
3	−1.87	−0.01	0.62	1429	−0.07	52.51	53.53
4	−3.64	−0.01	0.62	1389	−0.14	51.92	53.57
5	−1.55	−0.00	0.62	1363	−0.06	52.25	53.53
6	−3.02	−0.01	0.63	1311	−0.12	51.85	53.51
7	−2.64	−0.01	0.62	1407	−0.10	51.90	53.95
8	−1.50	−0.00	0.61	1397	−0.06	52.30	53.05
9	0.65	0.00	0.60	1425	0.02	51.89	52.84
High	3.73	0.01	0.59	1375	0.15	52.87	53.36

continued

Table 4.7 Continued

Portfolio	$\Delta\mu$	$\Delta\sigma$	p_A	NT	BETC	p_1	p_2
			Panel F: Volatility-sorted portfolios				
High	22.18	0.23	0.65	935	1.29	59.92	57.91
2	23.25	0.14	0.61	1097	1.15	57.69	56.13
3	20.25	0.11	0.61	1041	1.06	58.17	56.20
4	14.61	0.08	0.62	1053	0.75	58.25	56.70
5	12.91	0.07	0.63	1025	0.68	57.98	57.10
6	9.74	0.05	0.64	1047	0.51	57.37	57.05
7	9.20	0.05	0.65	1091	0.46	57.90	57.61
8	9.62	0.06	0.66	989	0.53	57.99	57.72
9	9.45	0.07	0.67	927	0.55	59.04	58.60
Low	8.92	0.08	0.68	831	0.58	61.28	59.99
			Panel G: Industry-sorted portfolios				
NoDur	2.05	0.01	0.63	1299	0.09	53.63	54.62
Dubrl	0.90	0.00	0.57	1391	0.04	51.79	51.15
Manuf	3.60	0.01	0.61	1315	0.15	52.89	53.22
Enrgy	−5.87	−0.01	0.59	1431	−0.22	51.87	52.46
HiTec	3.94	0.01	0.57	1407	0.15	52.44	52.39
Telcm	−3.93	−0.01	0.58	1461	−0.15	50.93	51.28
Shops	3.37	0.01	0.60	1313	0.14	53.43	53.47
Hlth	0.19	0.00	0.60	1331	0.01	52.20	53.13
Utils	3.85	0.01	0.61	1213	0.17	54.57	53.87
Other	5.09	0.02	0.61	1273	0.22	53.58	53.94

statistically insignificant estimates for the market timing regression coefficients indicating lack of evidence of market timing power of the MA strategy for those specific portfolios.

Figures 4.1 through 4.7 present the scatter plots of the daily returns of the MA strategy versus the daily returns of the BH strategy when short-selling is permitted for the size-sorted deciles, book-to-market sorted deciles, momentum-sorted deciles, short-term reversal deciles, long-term reversal deciles, volatility-sorted deciles, and industry portfolios, respectively.

Figures 4.8 through 4.14 present the cumulative investment performance of $1 invested on January 4, 1960, and continuously reinvested until December 31, 2013 the MA strategy (thin line) and the BH strategy (thick line) when short-selling is permitted for the size-sorted deciles,

Table 4.8 Market timing with short-selling

This table reports alphas, betas, and adjusted R^2 of the market-timing regressions of the MAP excess returns on the market factor using portfolios sorted by various characteristics when short-selling is permitted. The TM panel reports the results using the Treynor and Mazuy (1966) quadratic regression with the squared market factor (β_{m^2}) while the HM panel reports the results using the Merton and Henriksson (1981) regression with option-like returns on the market (γ_m). The sample period covers January 4, 1960 until December 31, 2013 with value-weighted portfolio returns. The length of the moving average window is 20 days. Newey and West (1987) standard errors with five lags are used in reporting statistical significance of a two-sided null hypothesis at the 1%, 5%, and 10% level given by a***, a**, and a*, respectively.

	TM				HM			
Portfolio	α	β_m	β_{m^2}	$\overline{R}^2$	α	β_m	γ_m	$\overline{R}^2$
				Panel A: Size-sorted portfolios				
Low	22.639***	−0.858***	0.026***	0.412	4.416	−0.715***	0.298***	0.412
2	20.781***	−1.117***	0.010***	0.454	10.220***	−1.040***	0.157***	0.455
3	16.576***	−1.142***	0.016***	0.484	4.592	−1.050***	0.191***	0.484
4	16.971***	−1.139***	0.008***	0.495	10.388***	−1.089***	0.104***	0.495
5	15.212***	−1.150***	0.003	0.511	11.658***	−1.124***	0.053*	0.511
6	13.560***	−1.103***	0.008***	0.529	8.789***	−1.065***	0.081***	0.529
7	12.885***	−1.126***	0.009***	0.540	9.000***	−1.093***	0.072***	0.540
8	9.954***	−1.161***	0.010***	0.560	5.355*	−1.121***	0.085***	0.560
9	2.853	−1.171***	0.019***	0.585	0.348	−1.135***	0.088***	0.583
High	−2.765	−1.216***	0.016***	0.590	−2.569	−1.201***	0.044*	0.588

continued

Table 4.8 Continued

Portfolio	TM				HM			
	α	β_m	β_{m^2}	$\overline{R}^2$	α	β_m	γ_m	$\overline{R}^2$
	Panel B: Book-to-market sorted portfolios							
Low	8.462***	−1.322***	0.005**	0.564	9.396***	−1.323***	0.004	0.563
2	4.641**	−1.194***	0.010***	0.565	4.754	−1.184***	0.029	0.564
3	3.246	−1.115***	0.017***	0.548	0.281	−1.078***	0.086***	0.546
4	2.266	−1.136***	0.025***	0.550	−2.023	−1.083***	0.124***	0.547
5	1.240	−1.102***	0.009***	0.520	3.794	−1.109***	−0.005	0.519
6	2.622	−1.063***	0.013***	0.519	2.039	−1.045***	0.045*	0.517
7	1.006	−1.017***	0.017***	0.502	−0.504	−0.989***	0.070***	0.500
8	0.382	−1.053***	0.019***	0.471	−0.859	−1.025***	0.071***	0.469
9	1.615	−1.071***	0.018***	0.484	−0.045	−1.041***	0.074***	0.482
High	6.208**	−1.155***	0.018***	0.459	3.001	−1.116***	0.091***	0.458

Panel C: Momentum-sorted portfolios								
Low	40.588***	−1.667***	−0.012***	0.451	43.077***	−1.695***	−0.067	0.451
2	23.291***	−1.398***	−0.009***	0.494	24.372***	−1.414***	−0.038	0.493
3	14.881***	−1.210***	−0.002	0.510	13.507***	−1.203***	0.010	0.510
4	7.051***	−1.176***	0.013***	0.534	3.544	−1.139***	0.082***	0.533
5	6.484***	−1.118***	−0.000	0.527	9.843***	−1.140***	−0.042*	0.527
6	2.736	−1.076***	0.011***	0.521	6.545**	−1.089***	−0.015	0.521
7	0.976	−1.085***	0.015***	0.533	0.643	−1.066***	0.050**	0.532
8	−3.474	−1.071***	0.024***	0.516	−6.452**	−1.027***	0.106***	0.513
9	−0.298	−1.182***	0.020***	0.527	−5.266	−1.129***	0.120***	0.525
High	4.077	−1.407***	0.020***	0.475	0.228	−1.361***	0.105***	0.474

continued

Table 4.8 Continued

	TM				HM			
Portfolio	α	β_m	β_{m^2}	$\overline{R}^2$	α	β_m	γ_m	$\overline{R}^2$
			Panel D: Short-term reversal sorted portfolios					
Low	1.569	−1.675***	−0.024***	0.392	20.770***	−1.821***	−0.305***	0.393
2	7.337**	−1.515***	−0.007**	0.495	15.154***	−1.571***	−0.114***	0.495
3	7.609***	−1.339***	−0.005*	0.509	14.200***	−1.386***	−0.094***	0.510
4	6.509***	−1.241***	0.012***	0.534	9.265***	−1.245***	0.002	0.533
5	3.965*	−1.185***	0.016***	0.552	2.887	−1.161***	0.060**	0.551
6	2.422	−1.113***	0.019***	0.541	0.871	−1.083***	0.074***	0.539
7	2.941	−1.107***	0.015***	0.559	1.451	−1.082***	0.062***	0.557
8	4.664**	−1.072***	0.016***	0.548	−2.393	−1.011***	0.133***	0.547
9	3.868*	−1.086***	0.014***	0.530	−1.714	−1.037***	0.108***	0.529
High	20.960***	−1.315***	0.014***	0.525	4.234	−1.195***	0.243***	0.527

Panel E: Long-term reversal sorted portfolios

Low	11.646***	−1.346***	0.023***	0.485	4.813	−1.279***	0.151***	0.484
2	2.872	−1.183***	0.020***	0.512	3.061	−1.163***	0.058**	0.509
3	−0.235	−1.078***	0.020***	0.506	−2.275	−1.044***	0.084***	0.504
4	−3.212	−1.062***	0.025***	0.507	−3.472	−1.034***	0.076***	0.503
5	−1.785	−1.053***	0.027***	0.512	−3.273	−1.015***	0.099***	0.507
6	0.642	−1.011***	0.010***	0.510	1.297	−1.005***	0.022	0.510
7	0.636	−1.036***	0.012***	0.517	−2.080	−1.005***	0.069***	0.516
8	1.304	−1.130***	0.017***	0.541	−3.162	−1.084***	0.103***	0.540
9	7.331***	−1.240***	0.003	0.537	5.248	−1.223***	0.035	0.537
High	14.936***	−1.533***	−0.008***	0.543	15.809***	−1.547***	−0.034	0.543

continued

Table 4.8 Continued

Portfolio	TM				HM			
	α	β_m	β_{m^2}	$\bar{R}^2$	α	β_m	γ_m	$\bar{R}^2$
Panel F: Volatility-sorted portfolios								
High	23.693***	−1.094***	0.021***	0.355	9.456**	−0.982***	0.235***	0.355
2	25.511***	−1.288***	0.023***	0.465	9.245**	−1.161***	0.264***	0.465
3	25.145***	−1.254***	0.011***	0.481	16.612***	−1.189***	0.136***	0.481
4	19.271***	−1.163***	0.010***	0.486	13.599***	−1.116***	0.098***	0.486
5	16.963***	−1.065***	0.010***	0.486	9.675***	−1.009***	0.117***	0.486
6	13.018***	−0.991***	0.011***	0.491	8.103**	−0.949***	0.093***	0.490
7	11.350***	−0.879***	0.013***	0.486	5.901**	−0.830***	0.105***	0.485
8	8.962***	−0.757***	0.022***	0.475	0.676	−0.682***	0.164***	0.472
9	7.901***	−0.597***	0.021***	0.423	−1.696	−0.514***	0.179***	0.419
Low	6.351***	−0.336***	0.019***	0.257	−1.233	−0.268***	0.148***	0.250

Panel G: Industry-sorted portfolios

NoDur	2.183	−0.856***	0.021***	0.440	−3.940	−0.796***	0.136***	0.437
Durbl	5.769*	−1.345***	0.014***	0.465	6.388	−1.334***	0.032	0.464
Manuf	5.968**	−1.174***	0.020***	0.546	3.318	−1.136***	0.090***	0.543
Enrgy	−1.124	−1.105***	0.008***	0.338	1.953	−1.115***	−0.014	0.338
HiTec	12.148***	−1.566***	0.005*	0.490	15.325***	−1.581***	−0.023	0.490
Telcm	1.919	−1.076***	0.003	0.413	7.061**	−1.106***	−0.054*	0.413
Shops	7.511***	−1.095***	0.010***	0.468	7.703**	−1.085***	0.028	0.467
Hlth	2.226	−1.040***	0.018***	0.416	1.054	−1.014***	0.066**	0.414
Utils	7.510***	−0.695***	0.002	0.300	6.363**	−0.686***	0.021	0.300
Other	10.398***	−1.292***	0.010***	0.537	5.999*	−1.254***	0.084***	0.537

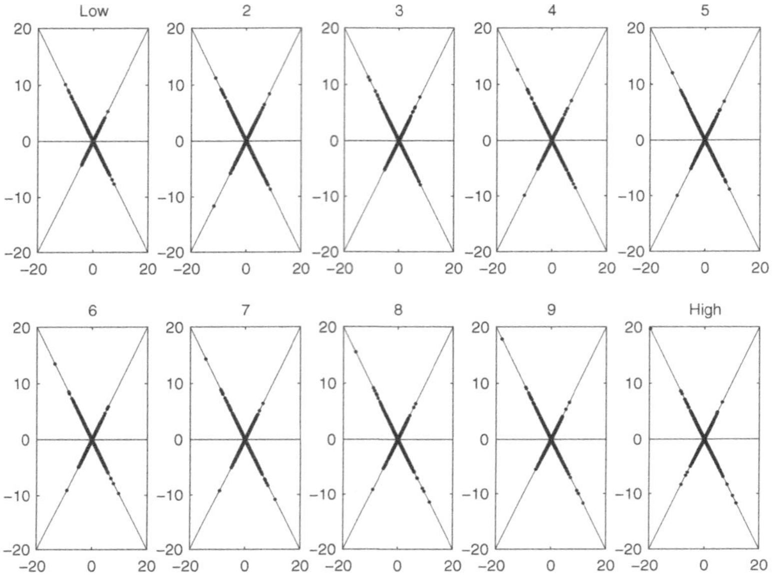

Figure 4.1 Scatter plot of BH versus MA returns with short-selling: Size decile portfolios.

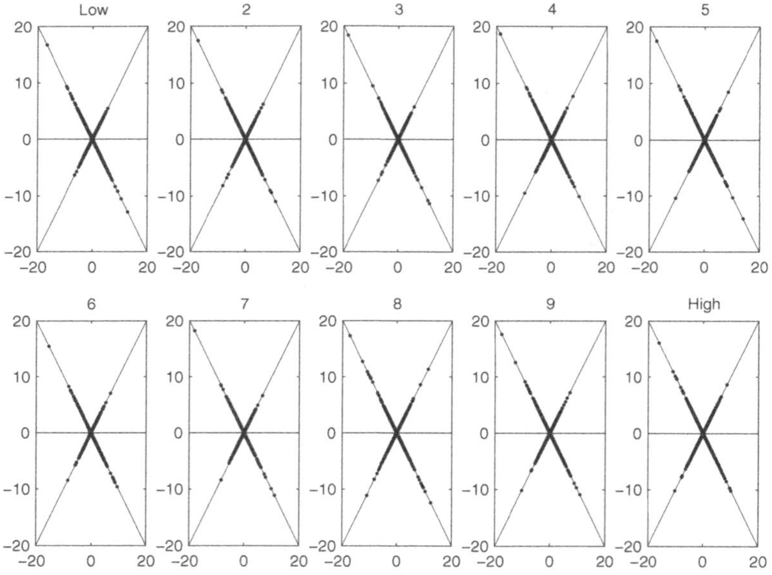

Figure 4.2 Scatter plot of BH versus MA returns with short-selling: Book-to-market decile portfolios.

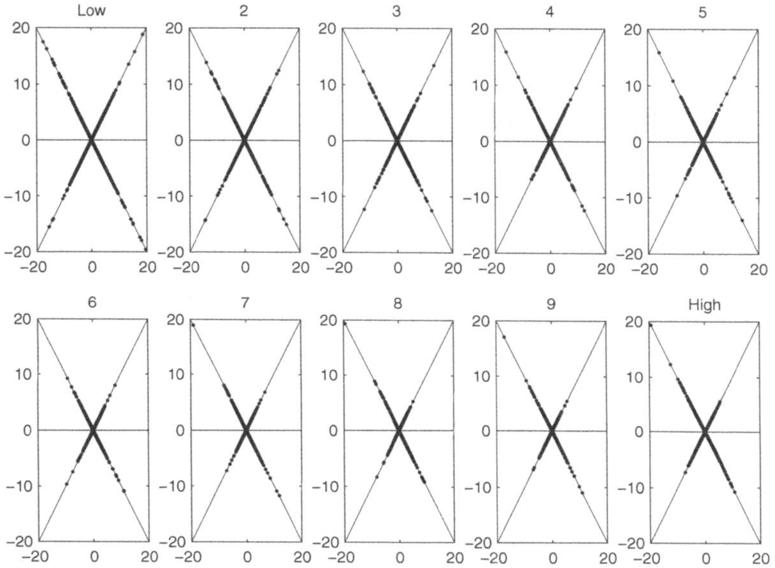

Figure 4.3 Scatter plot of BH versus MA returns with short-selling: Momentum decile portfolios.

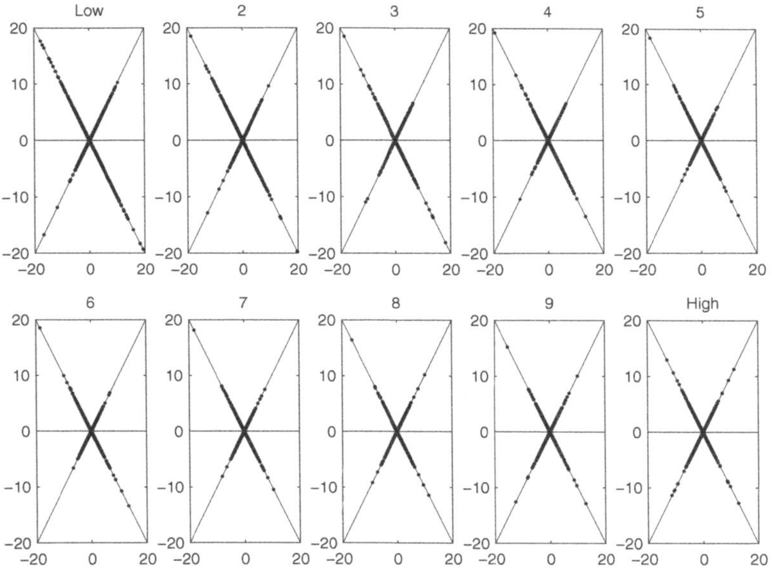

Figure 4.4 Scatter plot of BH versus MA returns with short-selling: Short-term reversal decile portfolios.

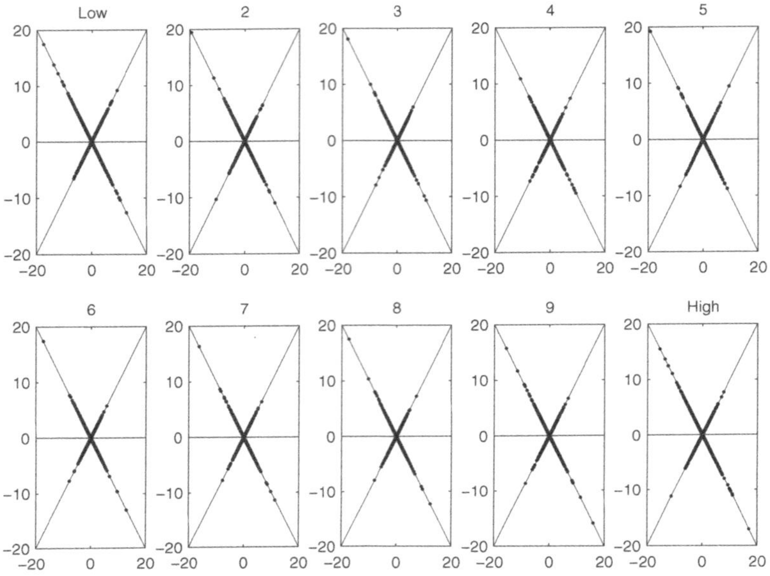

Figure 4.5 Scatter plot of BH versus MA returns with short-selling: Long-term reversal decile portfolios.

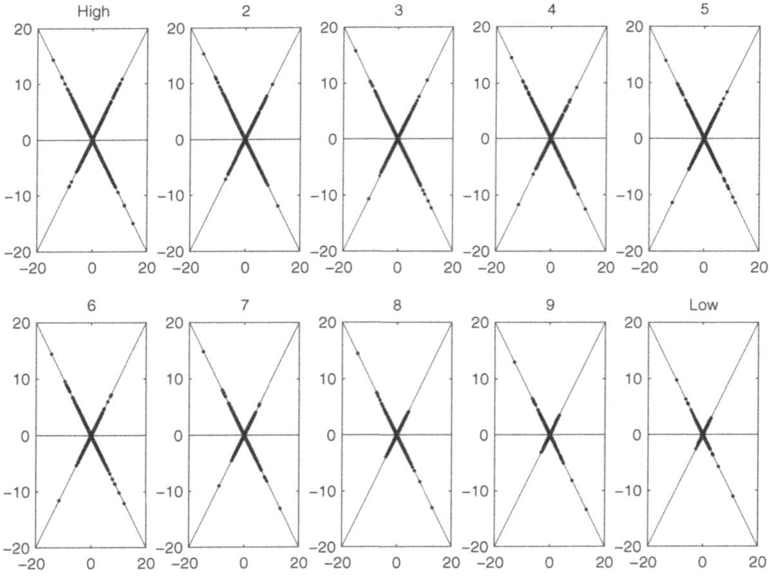

Figure 4.6 Scatter plot of BH versus MA returns with short-selling: Volatility decile portfolios.

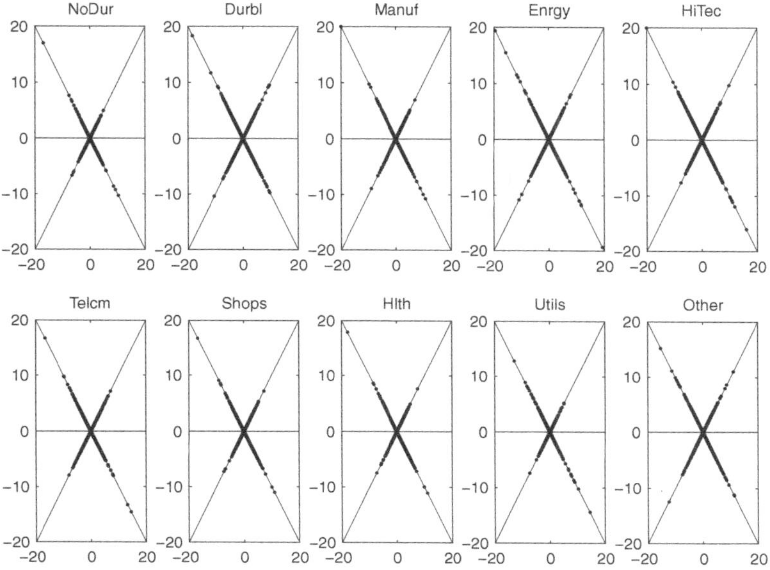

Figure 4.7 Scatter plot of BH versus MA returns with short-selling: Industry portfolios.

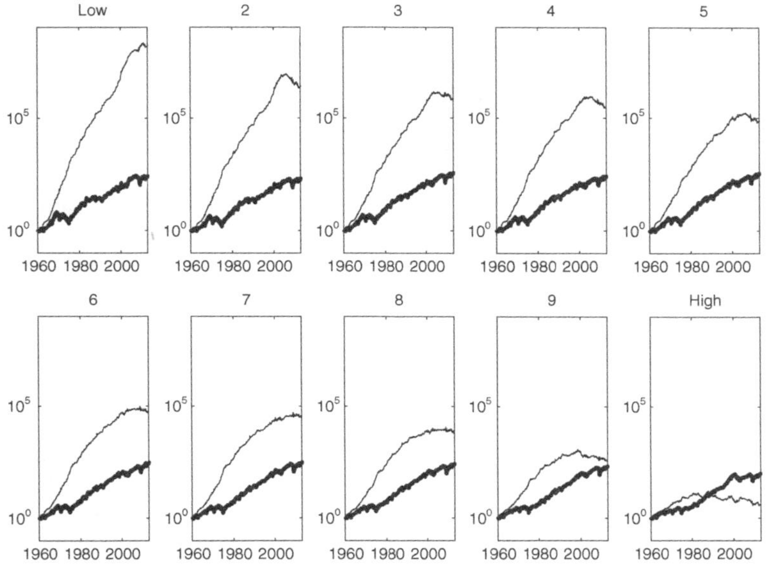

Figure 4.8 Cumulative returns of BH versus MA strategy with short-selling: Size decile portfolios.

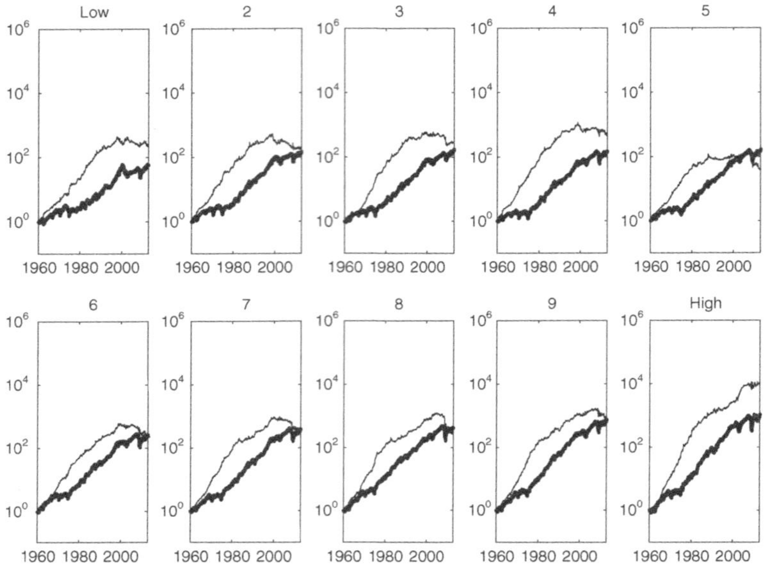

Figure 4.9 Cumulative returns of BH versus MA strategy with short-selling: Book-to-market decile portfolios.

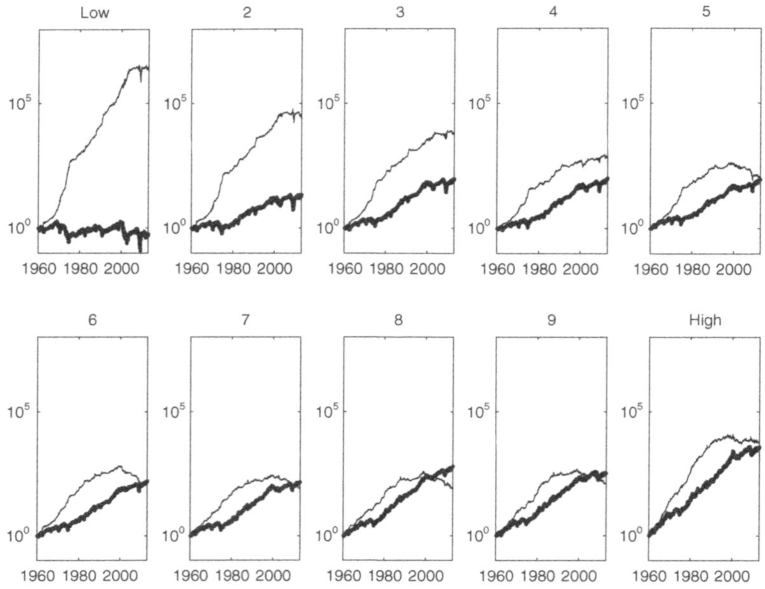

Figure 4.10 Cumulative returns of BH versus MA strategy with short-selling: Momentum decile portfolios.

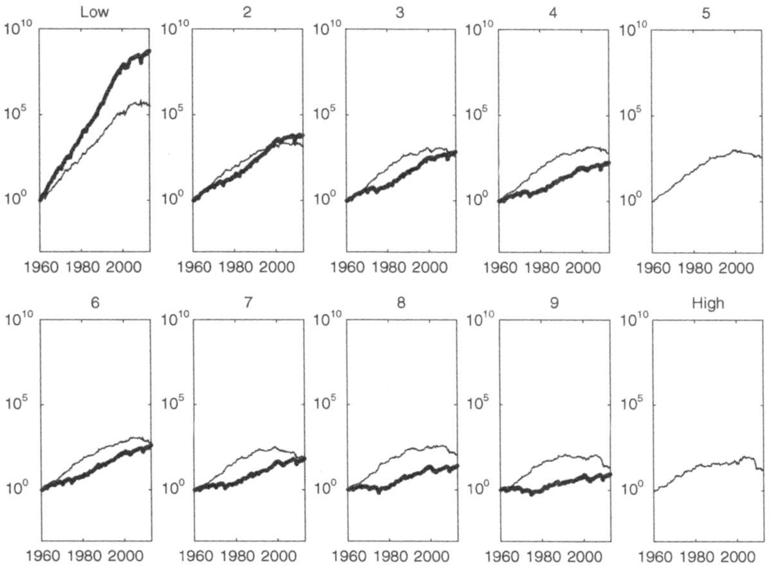

Figure 4.11 Cumulative Returns of BH versus MA strategy with short-selling: Short-term reversal decile portfolios.

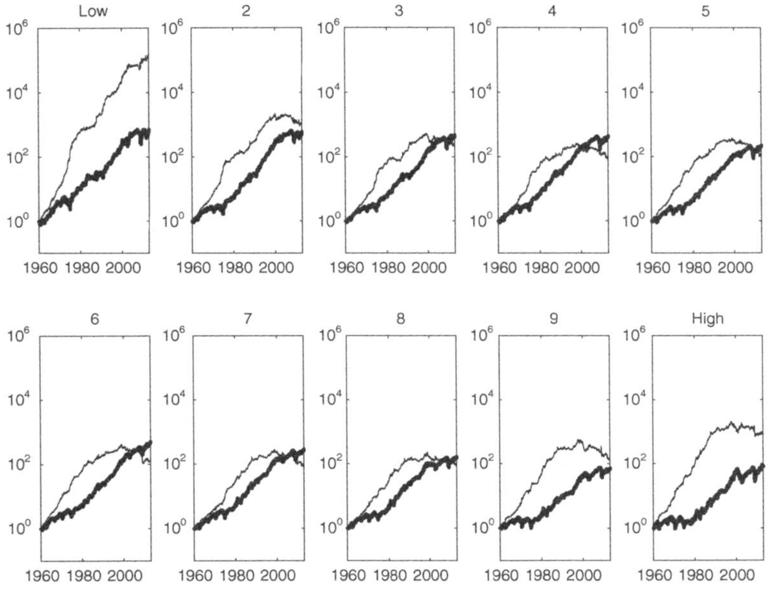

Figure 4.12 Cumulative returns of BH versus MA strategy with short-selling: Long-term reversal decile portfolios

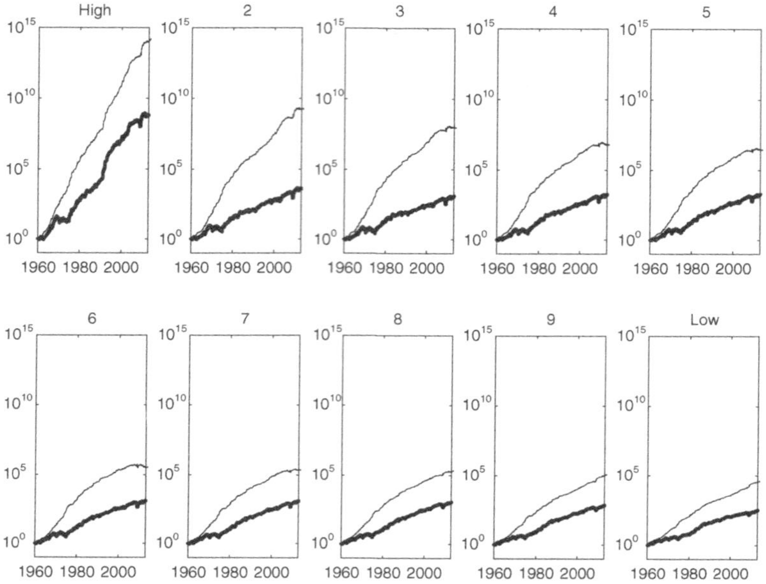

Figure 4.13 Cumulative returns of BH versus MA strategy with short-selling: Volatility decile portfolios.

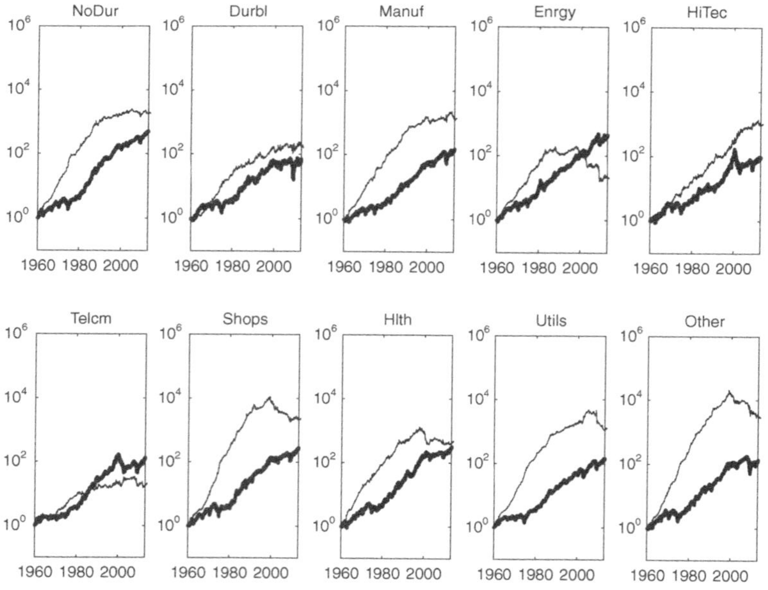

Figure 4.14 Cumulative returns of BH versus MA strategy with short-selling: Industry portfolios.

book-to-market sorted deciles, momentum-sorted deciles, short-term reversal deciles, long-term reversal deciles, volatility-sorted deciles, and industry portfolios, respectively.

4.3 Skipping a Period

In this case, the returns of the MA switching strategy can be expressed as follows:

$$\tilde{R}_{jt,L} = \begin{cases} R_{jt}, & \text{if } P_{jt-2} > A_{jt-2,L}, \\ r_{ft}, & \text{otherwise,} \end{cases} \tag{4.3}$$

in the absence of any transaction costs imposed on the switches. The alternative specification for the case of positive one-way transaction cost of τ leads to the following four cases in the post-transaction cost returns:

$$\tilde{R}_{jt,L} = \begin{cases} R_{jt}, & \text{if } P_{jt-2} > A_{jt-2,L} \text{ and } P_{jt-3} > A_{jt-3,L}, \\ R_{jt} - \tau, & \text{if } P_{jt-2} > A_{jt-2,L} \text{ and } P_{jt-3} < A_{jt-3,L}, \\ r_{ft}, & \text{if } P_{jt-2} < A_{jt-2,L} \text{ and } P_{jt-3} < A_{jt-3,L}, \\ r_{ft} - \tau, & \text{if } P_{jt-2} < A_{jt-2,L} \text{ and } P_{jt-3} > A_{jt-3,L}. \end{cases} \tag{4.4}$$

Table 4.9 presents the summary statistics for the performance of the MA, BH, and MAP returns when we skip a day before we trade. It is instructive to compare these to the results for the baseline implementation reported in Table 2.1. As expected, delaying trading by even one day reduces the average returns of the MA strategy relative to the BH strategy. This leads to the underperformance of the MA strategy relative to the BH strategy in more cases than it does using the baseline implementation.

Table 4.10 reports the parameter estimates for the CAPM model, the Fama-French three-factor model, and the Carhart four-factor model. Once again, the number and magnitude of positive and statistically significant abnormal returns are reduced when compared to the baseline case investigated in the previous section and chapters. The level of annualized αs can be as high as 16% per year for the smallest capitalization decile when using the four-factor model. This leads to the conclusion that skipping one day before trading results in reduced performance of the MA strategy compared to the BH strategy.

Table 4.9 Portfolio performance with skipping a day

This table reports summary statistics for the respective buy and hold (BH) portfolio returns, the MA switching strategy portfolio returns with skipping a day before trading and the excess return of MA over BH (MAP) using sets of ten portfolios sorted by various characteristics. The sample period covers January 4, 1960 until December 31, 2013 with value-weighted portfolio returns. μ is the annualized average return, σ is the annualized standard deviation of returns, s is the annualized skewness, k is the annualized kurtosis, and SR is the annualized Sharpe ratio. The length of the MA window is 20 days.

	BH portfolios					**MA portfolios**					**MAP portfolios**				
Portfolio	μ	σ	s	k	SR	μ	σ	s	k	SR	μ	σ	s	k	SR
						Panel A: Size-sorted portfolios									
Low	11.88	13.58	−0.90	14.81	0.52	22.43	8.50	−0.46	12.70	2.08	10.54	10.53	1.49	34.96	1.00
2	11.86	16.72	−0.46	13.85	0.52	18.81	10.60	−0.21	15.83	2.08	6.95	12.90	0.73	31.62	1.00
3	12.96	16.60	−0.49	12.02	0.42	17.63	10.46	−0.23	13.27	1.32	4.67	12.87	0.75	27.21	0.54
4	12.18	16.33	−0.48	12.48	0.49	16.56	10.36	−0.19	12.29	1.23	4.37	12.61	0.79	29.27	0.36
5	12.68	16.23	−0.47	11.96	0.45	15.30	10.31	−0.29	12.06	1.14	2.62	12.52	0.71	28.01	0.35
6	12.26	15.33	−0.53	13.43	0.49	14.22	9.80	−0.26	11.28	1.02	1.96	11.79	0.87	32.83	0.21
7	12.29	15.47	−0.55	15.54	0.49	13.19	9.80	−0.33	12.01	0.96	0.90	11.98	0.88	37.76	0.17
8	12.01	15.70	−0.51	16.62	0.49	12.37	10.06	−0.22	11.94	0.86	0.36	12.05	0.87	41.87	0.08
9	11.57	15.46	−0.59	20.90	0.46	10.05	9.75	−0.22	11.85	0.75	−1.52	12.00	1.02	52.24	0.03
High	10.23	16.00	−0.50	21.64	0.44	6.25	10.25	−0.28	13.46	0.54	−3.97	12.29	0.86	55.66	−0.13

Panel B: Book-to-market sorted portfolios

Low	9.49	17.53	−0.18	13.47	0.27	8.59	10.96	−0.16	9.82	0.35	−0.90	13.68	0.22	32.27	−0.07
2	11.03	16.07	−0.39	16.37	0.27	8.99	10.25	−0.20	11.07	0.35	−2.05	12.38	0.64	41.26	−0.07
3	11.15	15.42	−0.52	20.70	0.39	9.10	9.99	−0.30	11.59	0.41	−2.04	11.76	0.88	55.21	−0.17
4	10.95	15.80	−0.71	20.94	0.41	10.11	10.12	−0.05	13.30	0.43	−0.84	12.14	1.43	53.59	−0.17
5	11.13	15.62	−0.45	23.02	0.39	9.55	10.00	−0.38	19.23	0.53	−1.58	12.00	0.66	56.74	−0.07
6	11.89	15.14	−0.45	16.96	0.41	9.90	9.72	−0.21	12.78	0.48	−1.99	11.61	0.74	42.66	−0.13
7	12.55	14.89	−0.64	22.73	0.47	10.55	9.76	−0.41	14.80	0.53	−2.00	11.26	1.08	61.17	−0.17
8	12.98	15.82	−0.60	30.16	0.52	11.07	10.34	−0.24	35.79	0.59	−1.91	11.97	1.09	71.74	−0.18
9	13.90	15.99	−0.68	20.87	0.52	13.55	10.43	−0.21	14.14	0.61	−0.36	12.12	1.27	55.18	−0.16
High	15.01	17.78	−0.46	15.88	0.57	14.70	11.70	−0.16	15.22	0.84	−0.31	13.39	0.80	40.32	−0.03

continued

Table 4.9 Continued

Portfolio	BH portfolios					MA portfolios					MAP portfolios				
	μ	σ	s	k	SR	μ	σ	s	k	SR	μ	σ	s	k	SR
Panel C: Momentum-sorted portfolios															
Low	2.59	24.93	0.41	25.70	−0.09	14.00	15.67	1.51	56.95	0.59	11.41	19.38	−0.10	46.00	0.59
2	8.03	20.35	0.15	22.17	−0.09	12.15	12.92	0.27	35.19	0.59	4.11	15.72	−0.26	46.27	0.59
3	10.29	17.59	0.04	17.98	0.16	11.50	11.60	0.16	32.62	0.57	1.21	13.22	−0.08	37.01	0.26
4	10.22	16.61	−0.15	18.81	0.31	10.61	10.76	0.26	19.14	0.58	0.40	12.66	0.41	45.74	0.09
5	10.09	15.89	0.01	21.65	0.33	9.12	10.44	0.18	25.44	0.54	−0.97	11.99	0.01	52.27	0.03
6	10.99	15.35	−0.58	26.71	0.33	9.32	10.03	−0.23	18.15	0.42	−1.67	11.63	1.09	71.05	−0.08
7	10.92	15.19	−0.55	22.88	0.40	8.80	9.80	−0.32	12.15	0.45	−2.12	11.62	0.94	60.72	−0.14
8	13.57	15.49	−0.73	21.21	0.40	9.91	10.24	−0.40	11.23	0.41	−3.66	11.64	1.32	59.75	−0.18
9	12.81	16.47	−0.54	15.10	0.57	9.18	10.55	−0.42	9.84	0.50	−3.63	12.66	0.83	38.48	−0.31
High	17.85	20.37	−0.50	12.74	0.49	13.24	12.90	−0.72	9.18	0.42	−4.61	15.78	0.54	31.15	−0.29

Panel D: Short-term Reversal sorted portfolios

Low	41.02	25.55	0.87	29.76	1.42	28.43	14.93	0.27	17.60	1.58	−12.59	20.79	−1.81	63.73	−0.61
2	18.83	20.36	0.33	27.47	1.42	13.92	11.46	0.44	18.21	1.58	−4.91	16.85	−0.59	54.86	−0.61
3	14.32	17.97	0.06	26.89	0.69	11.68	10.52	−0.02	19.37	0.80	−2.64	14.58	−0.24	56.93	−0.29
4	11.57	16.65	−0.38	23.86	0.53	10.20	10.06	−0.35	18.41	0.66	−1.37	13.28	0.51	53.00	−0.18
5	11.73	15.81	−0.47	22.77	0.41	10.40	9.82	−0.19	12.56	0.54	−1.33	12.39	0.77	55.26	−0.10
6	12.78	15.57	−0.45	22.73	0.44	11.25	10.09	−0.10	12.41	0.57	−1.53	11.87	0.81	60.81	−0.11
7	9.63	15.45	−0.55	19.99	0.51	9.47	10.23	−0.15	13.53	0.64	−0.16	11.58	1.10	55.00	−0.13
8	7.93	15.66	−0.42	16.68	0.31	8.03	10.86	−0.28	19.62	0.46	0.10	11.29	0.80	45.02	−0.01
9	6.01	16.52	−0.44	15.75	0.20	4.19	11.69	−0.75	22.77	0.30	−1.81	11.67	0.46	40.31	0.01
High	−9.59	19.08	−0.55	11.88	0.07	−2.90	12.91	−0.95	25.15	−0.05	6.68	14.07	0.80	22.56	−0.16

continued

Table 4.9 Continued

Portfolio	BH portfolios					MA portfolios					MAP portfolios				
	μ	σ	s	k	SR	μ	σ	s	k	SR	μ	σ	s	k	SR
Panel E: Long-term Reversal sorted portfolios															
Low	14.73	20.02	−0.38	14.61	0.50	16.95	13.03	0.08	12.82	0.93	2.22	15.19	0.77	36.93	0.15
2	13.72	17.10	−0.61	18.75	0.50	12.22	11.15	−0.15	11.81	0.93	−1.50	12.97	1.16	50.07	0.15
3	13.09	15.94	−0.58	18.32	0.52	11.37	10.63	−0.20	12.44	0.67	−1.72	11.89	1.11	51.17	−0.12
4	12.96	15.66	−0.79	25.23	0.52	10.51	10.29	−0.18	12.54	0.62	−2.45	11.81	1.60	70.56	−0.14
5	11.63	15.50	−0.71	22.89	0.52	10.08	10.20	−0.06	16.03	0.56	−1.54	11.67	1.50	61.64	−0.21
6	12.91	14.73	−0.36	21.45	0.44	10.61	9.78	−0.09	11.34	0.52	−2.31	11.02	0.65	61.45	−0.13
7	12.10	15.03	−0.39	17.81	0.55	9.41	10.00	−0.30	12.70	0.60	−2.69	11.23	0.60	49.12	−0.21
8	11.03	15.60	−0.46	19.22	0.49	9.03	9.95	−0.25	13.16	0.46	−2.01	12.02	0.78	48.35	−0.24
9	9.75	16.87	−0.19	18.39	0.40	7.75	10.66	−0.37	13.79	0.43	−2.00	13.08	0.13	44.84	−0.17
High	10.66	20.33	−0.16	14.88	0.29	9.28	12.48	−0.43	13.96	0.28	−1.37	16.06	0.06	33.14	−0.15

Panel F: Volatility-sorted portfolios

High	40.10	20.27	0.10	14.40	1.74	48.88	13.99	1.11	17.12	3.15	8.78	14.56	0.22	38.88	0.60
2	17.80	19.30	−0.31	14.52	1.74	26.23	12.24	0.31	16.55	3.15	8.42	14.87	0.65	33.24	0.60
3	15.22	18.25	−0.38	16.39	0.67	22.30	11.52	0.15	16.83	1.75	7.07	14.13	0.71	37.94	0.57
4	15.74	16.89	−0.47	18.77	0.57	20.57	10.65	0.11	17.26	1.52	4.83	13.09	0.88	44.14	0.50
5	15.68	15.52	−0.55	20.47	0.65	19.66	9.77	0.03	17.03	1.48	3.98	12.03	0.98	48.80	0.37
6	14.82	14.38	−0.67	24.45	0.70	18.07	9.03	−0.00	17.21	1.52	3.25	11.18	1.21	59.19	0.33
7	14.40	12.78	−0.66	28.32	0.70	16.82	8.01	−0.18	13.69	1.47	2.42	9.95	1.08	70.97	0.29
8	13.68	11.15	−0.78	36.32	0.75	16.80	6.94	−0.17	11.50	1.50	3.11	8.72	1.31	92.31	0.24
9	12.53	9.27	−0.58	54.72	0.80	15.55	5.66	−0.20	12.00	1.73	3.02	7.33	0.84	135.61	0.36
Low	10.81	6.86	−0.47	71.59	0.84	14.16	4.17	0.07	14.58	1.90	3.36	5.43	0.70	177.24	0.41

continued

Table 4.9 Continued

Portfolio	BH portfolios					MA portfolios					MAP portfolios				
	μ	σ	s	k	SR	μ	σ	s	k	SR	μ	σ	s	k	SR
Panel G: Industry-sorted portfolios															
NoDur	12.80	13.63	−0.62	22.17	0.59	12.92	9.00	−0.24	10.91	0.90	0.12	10.23	1.16	63.01	0.01
Durbl	10.57	20.63	−0.23	11.77	0.59	9.17	13.70	0.15	14.29	0.90	−1.40	15.43	0.58	28.71	0.01
Manuf	11.07	16.28	−0.68	21.25	0.28	9.85	10.48	−0.19	12.26	0.32	−1.22	12.46	1.30	55.65	−0.09
Enrgy	13.49	19.99	−0.20	19.86	0.39	8.07	13.58	−0.28	17.64	0.48	−5.41	14.68	0.18	55.42	−0.10
HiTec	11.37	22.42	0.02	12.00	0.44	11.17	14.15	0.04	10.76	0.24	−0.20	17.39	−0.10	28.45	−0.37
Telcm	10.94	17.36	−0.03	17.12	0.29	7.92	11.50	−0.06	14.16	0.45	−3.02	13.00	−0.05	45.75	−0.01
Shops	12.23	16.54	−0.28	14.21	0.36	11.88	10.81	−0.13	11.54	0.27	−0.34	12.53	0.45	36.79	−0.23
Hlth	12.52	16.83	−0.40	14.37	0.45	9.87	11.15	−0.35	10.45	0.66	−2.64	12.61	0.60	39.11	−0.03
Utils	10.18	12.93	−0.03	28.88	0.46	12.00	8.61	−1.21	42.54	0.46	1.83	9.64	−0.95	66.67	−0.21
Other	11.15	17.73	−0.26	18.07	0.42	12.57	11.27	0.04	22.60	0.84	1.42	13.70	0.48	40.38	0.19

Table 4.10 Factor loadings with skipping a day

This table reports alphas, betas, and adjusted R^2 of the regressions of the MAP excess returns on the market factor, the Fama-French three-factors, and the Carhart four-factors using portfolios sorted by various characteristics. The alphas are annualized and in percent. The sample period covers January 4, 1960 until December 31, 2013 with daily value-weighted portfolio returns. The length of the MA window is 20 days. Newey and West (1987) standard errors with five lags are used in reporting statistical significance of a two-sided null hypothesis at the 1%, 5%, and 10% level given by a***, a**, and a*, respectively.

	CAPM			Fama-French					Carhart					
Portfolio	α	β_m	$\bar{R}^2$	α	β_m	β_s	β_h	$\bar{R}^2$	α	β_m	β_s	β_h	β_u	$\bar{R}^2$
Panel A: Size-sorted portfolios														
Low	13.130***	−0.429***	0.399	14.801***	−0.485***	−0.511***	−0.138***	0.543	15.568***	−0.495***	−0.509***	−0.164***	−0.072***	0.548
2	10.323***	−0.559***	0.452	12.289***	−0.625***	−0.586***	−0.168***	0.579	12.963***	−0.634***	−0.584***	−0.190***	−0.063***	0.582
3	8.152***	−0.576***	0.482	9.853***	−0.634***	−0.550***	−0.130***	0.594	10.554***	−0.643***	−0.549***	−0.153***	−0.065***	0.597
4	7.815***	−0.570***	0.492	9.260***	−0.620***	−0.499***	−0.100***	0.587	10.020***	−0.630***	−0.497***	−0.125***	−0.071***	0.590
5	6.097***	−0.577***	0.510	7.272***	−0.618***	−0.444***	−0.068***	0.586	7.885***	−0.626***	−0.442***	−0.089***	−0.057***	0.588
6	5.283***	−0.550***	0.524	6.030***	−0.577***	−0.310***	−0.034***	0.565	6.816***	−0.587***	−0.308***	−0.060***	−0.073***	0.569
7	4.326***	−0.567***	0.540	5.130***	−0.595***	−0.254***	−0.064***	0.567	5.937***	−0.605***	−0.252***	−0.091***	−0.075***	0.572
8	3.846***	−0.577***	0.551	4.546***	−0.599***	−0.184***	−0.068***	0.566	5.300***	−0.609***	−0.182***	−0.093***	−0.070***	0.569
9	2.039*	−0.590***	0.581	2.505**	−0.603***	−0.057***	−0.068***	0.584	3.356***	−0.614***	−0.055***	−0.096***	−0.079***	0.588
High	−0.333	−0.603***	0.579	−0.878	−0.585***	0.154***	0.049***	0.589	−0.419	−0.591***	0.155***	0.034***	−0.043***	0.590

continued

Table 4.10 Continued

	CAPM			Fama-French					Carhart					
Portfolio	α	β_m	$\bar{R}^2$	α	β_m	β_s	β_h	$\bar{R}^2$	α	β_m	β_s	β_h	β_u	$\bar{R}^2$
Panel B: Book-to-market sorted portfolios														
Low	3.099**	−0.662***	0.564	1.344	−0.615***	0.039***	0.314***	0.591	1.687	−0.619***	0.040***	0.303***	−0.032***	0.591
2	1.569	−0.599***	0.564	1.085	−0.586***	0.025***	0.082***	0.566	1.759	−0.595***	0.027***	0.059***	−0.063***	0.569
3	1.314	−0.556***	0.539	1.063	−0.549***	0.038***	0.034***	0.540	1.950*	−0.560***	0.040***	0.004	−0.083***	0.545
4	2.670**	−0.581***	0.552	3.417***	−0.601***	−0.001	−0.139***	0.559	3.886***	−0.607***	−0.000	−0.155***	−0.044***	0.560
5	1.797	−0.559***	0.522	2.791**	−0.585***	0.004	−0.187***	0.534	3.301***	−0.591***	0.005	−0.204***	−0.048***	0.536
6	1.276	−0.540***	0.521	2.423**	−0.572***	−0.060***	−0.194***	0.536	2.820**	−0.577***	−0.059***	−0.207***	−0.037***	0.537
7	1.082	−0.511***	0.496	2.465**	−0.548***	−0.028***	−0.249***	0.521	3.161***	−0.557***	−0.026***	−0.272***	−0.065***	0.524
8	1.292	−0.531***	0.473	3.600***	−0.593***	−0.061***	−0.410***	0.533	3.903***	−0.597***	−0.060***	−0.420***	−0.028***	0.534
9	2.921**	−0.543***	0.483	5.126***	−0.602***	−0.057***	−0.392***	0.536	5.928***	−0.613***	−0.054***	−0.419***	−0.075***	0.541
High	3.217**	−0.584***	0.458	5.922***	−0.659***	−0.179***	−0.444***	0.521	6.723***	−0.670***	−0.177***	−0.471***	−0.075***	0.524

Panel C: Momentum-sorted portfolios

Low	16.499***	−0.843***	0.456	18.174***	−0.894***	−0.272***	−0.220***	0.473	13.224***	−0.830***	−0.285***	−0.055***	0.462***	0.534
2	8.431***	−0.715***	0.498	9.596***	−0.748***	−0.093***	−0.186***	0.507	5.949***	−0.701***	−0.103***	−0.064***	0.341***	0.558
3	4.877***	−0.607***	0.509	5.674***	−0.628***	0.009	−0.152***	0.516	3.550***	−0.601***	0.004	−0.081***	0.198***	0.540
4	3.982***	−0.594***	0.530	4.653***	−0.611***	0.024**	−0.134***	0.536	3.664***	−0.598***	0.022**	−0.100***	0.092***	0.542
5	2.404**	−0.559***	0.523	2.817**	−0.570***	0.009	−0.080***	0.525	2.597**	−0.567***	0.008	−0.073***	0.021***	0.525
6	1.586	−0.539***	0.517	2.039*	−0.550***	0.021**	−0.092***	0.521	2.604**	−0.558***	0.023**	−0.111***	−0.053***	0.523
7	1.175	−0.546***	0.531	1.497	−0.554***	0.020**	−0.067***	0.533	2.654**	−0.569***	0.023***	−0.106***	−0.108***	0.542
8	−0.429	−0.535***	0.509	−0.146	−0.542***	0.019**	−0.059***	0.510	1.878*	−0.568***	0.024***	−0.127***	−0.189***	0.539
9	−0.069	−0.590***	0.524	0.174	−0.598***	−0.062***	−0.024**	0.525	2.764**	−0.632***	−0.055***	−0.111***	−0.242***	0.565
High	−0.419	−0.694***	0.465	−0.895	−0.687***	−0.227***	0.166***	0.485	3.407**	−0.742***	−0.216***	0.022	−0.402***	0.555

continued

Table 4.10 Continued

Portfolio	CAPM			Fama-French					Carhart					
	α	β_m	$\overline{R}^2$	α	β_m	β_s	β_h	$\overline{R}^2$	α	β_m	β_s	β_h	β_u	$\overline{R}^2$
Panel D: Short-term reversal sorted portfolios														
Low	−7.519***	−0.840***	0.392	−6.989***	−0.859***	−0.238***	−0.018	0.400	−8.161***	−0.844***	−0.241***	0.021	0.109***	0.403
2	−0.271	−0.768***	0.500	0.162	−0.782***	−0.103***	−0.046***	0.502	−0.042	−0.779***	−0.104***	−0.039***	0.019*	0.502
3	1.419	−0.673***	0.512	1.815	−0.684***	−0.060***	−0.054***	0.514	2.366*	−0.692***	−0.058***	−0.072***	−0.051***	0.515
4	2.382*	−0.621***	0.527	2.570**	−0.626***	−0.012	−0.031***	0.527	3.303***	−0.636***	−0.010	−0.056***	−0.068***	0.530
5	2.208*	−0.586***	0.538	2.477**	−0.593***	−0.011	−0.047***	0.538	3.154***	−0.602***	−0.009	−0.069***	−0.063***	0.541
6	1.854*	−0.560***	0.536	2.224**	−0.570***	−0.005	−0.067***	0.538	2.907***	−0.579***	−0.003	−0.090***	−0.064***	0.541
7	3.205***	−0.558***	0.558	3.546***	−0.567***	0.001	−0.064***	0.560	4.137***	−0.574***	0.002	−0.084***	−0.055***	0.562
8	3.335***	−0.535***	0.541	3.500***	−0.539***	0.004	−0.032***	0.541	4.280***	−0.549***	0.006	−0.058***	−0.073***	0.546
9	1.473	−0.544***	0.523	1.902*	−0.556***	−0.032***	−0.069***	0.525	2.188*	−0.560***	−0.032***	−0.079***	−0.027***	0.526
High	10.618***	−0.652***	0.517	11.194***	−0.671***	−0.177***	−0.047***	0.526	11.165***	−0.671***	−0.177***	−0.046***	0.003	0.526

Panel E: Long-term reversal sorted portfolios

Low	6.321***	−0.679***	0.481	8.103***	−0.734***	−0.340***	−0.217***	0.519	9.432***	−0.752***	−0.336***	−0.262***	−0.124***	0.526
2	2.099	−0.596***	0.508	3.352***	−0.632***	−0.152***	−0.182***	0.525	4.653***	−0.649***	−0.149***	−0.226***	−0.122***	0.534
3	1.551	−0.542***	0.500	2.515**	−0.569***	−0.056***	−0.161***	0.511	3.594***	−0.583***	−0.053***	−0.197***	−0.101***	0.518
4	0.832	−0.543***	0.509	1.997*	−0.574***	−0.011	−0.214***	0.526	2.688**	−0.583***	−0.009	−0.237***	−0.065***	0.529
5	1.707	−0.539***	0.512	2.732**	−0.565***	0.013	−0.196***	0.527	3.320***	−0.573***	0.015	−0.215***	−0.055***	0.529
6	0.740	−0.504***	0.505	1.300	−0.518***	0.044***	−0.120***	0.512	2.115**	−0.529***	0.046***	−0.147***	−0.076***	0.518
7	0.438	−0.518***	0.511	0.671	−0.523***	0.044***	−0.058***	0.514	1.368	−0.532***	0.046***	−0.082***	−0.065***	0.517
8	1.415	−0.567***	0.536	1.401	−0.565***	0.059***	−0.018*	0.537	2.321**	−0.577***	0.062***	−0.048***	−0.086***	0.543
9	1.701	−0.612***	0.528	1.458	−0.605***	0.052***	0.028**	0.529	2.115*	−0.613***	0.054***	0.006	−0.061***	0.531
High	3.232**	−0.763***	0.543	1.901	−0.729***	−0.055***	0.268***	0.559	2.055	−0.731***	−0.055***	0.262***	−0.014*	0.559

continued

Table 4.10 Continued

	CAPM			Fama-French					Carhart					
Portfolio	α	β_m	$\bar{R}^2$	α	β_m	β_s	β_b	$\bar{R}^2$	α	β_m	β_s	β_b	β_u	$\bar{R}^2$
							Panel F: Volatility-sorted portfolios							
High	12.080***	−0.547***	0.339	13.862***	−0.606***	−0.528***	−0.153***	0.421	14.236***	−0.611***	−0.527***	−0.166***	−0.035***	0.421
2	12.346***	−0.649***	0.459	14.082***	−0.707***	−0.498***	−0.155***	0.529	14.272***	−0.710***	−0.497***	−0.161***	−0.018*	0.529
3	10.879***	−0.630***	0.478	12.692***	−0.688***	−0.444***	−0.188***	0.543	13.081***	−0.694***	−0.443***	−0.201***	−0.036***	0.543
4	8.367***	−0.586***	0.483	10.235***	−0.645***	−0.378***	−0.220***	0.543	10.503***	−0.648***	−0.377***	−0.229***	−0.025***	0.544
5	7.250***	−0.541***	0.487	9.015***	−0.595***	−0.301***	−0.227***	0.538	9.304***	−0.599***	−0.301***	−0.237***	−0.027***	0.539
6	6.312***	−0.507***	0.496	7.937***	−0.556***	−0.232***	−0.224***	0.538	8.207***	−0.559***	−0.231***	−0.234***	−0.025***	0.538
7	5.129***	−0.448***	0.488	6.329***	−0.484***	−0.161***	−0.169***	0.516	6.737***	−0.489***	−0.160***	−0.183***	−0.038***	0.518
8	5.439***	−0.385***	0.470	6.341***	−0.411***	−0.104***	−0.133***	0.488	6.909***	−0.419***	−0.103***	−0.152***	−0.053***	0.492
9	4.853***	−0.303***	0.412	5.631***	−0.325***	−0.067***	−0.123***	0.430	6.123***	−0.332***	−0.066***	−0.139***	−0.046***	0.434
Low	4.399***	−0.173***	0.244	4.676***	−0.181***	−0.017***	−0.046***	0.248	5.154***	−0.187***	−0.016***	−0.062***	−0.045***	0.255

Panel G: Industry-sorted portfolios

NoDur	2.766**	−0.438***	0.441	2.669**	−0.434***	0.065***	−0.004	0.444	3.677***	−0.447***	0.068***	−0.038***	−0.094***	0.453
Durbl	2.723*	−0.683***	0.472	3.916**	−0.715***	−0.046***	−0.207***	0.481	4.249***	−0.720***	−0.045***	−0.218***	−0.031***	0.482
Manuf	2.356**	−0.592***	0.544	2.770**	−0.603***	−0.015	−0.072***	0.545	3.624***	−0.615***	−0.013	−0.101***	−0.080***	0.550
Enrgy	−2.141	−0.541***	0.328	−1.626	−0.552***	0.131***	−0.141***	0.338	−0.206	−0.570***	0.135***	−0.188***	−0.133***	0.347
HiTec	4.528***	−0.783***	0.488	2.096	−0.720***	−0.049***	0.471***	0.527	1.885	−0.717***	−0.050***	0.478***	0.020**	0.527
Telcm	0.182	−0.531***	0.401	0.540	−0.537***	0.121***	−0.108***	0.410	0.519	−0.537***	0.121***	−0.107***	0.002	0.410
Shops	2.992**	−0.553***	0.468	2.719**	−0.546***	−0.019*	0.057***	0.470	3.336**	−0.554***	−0.017	0.037***	−0.058***	0.472
Hlth	0.483	−0.518***	0.405	−0.339	−0.494***	0.076***	0.127***	0.412	0.571	−0.506***	0.079***	0.097***	−0.085***	0.417
Utils	3.811***	−0.328***	0.280	4.899***	−0.356***	0.048***	−0.219***	0.309	5.483***	−0.364***	0.049***	−0.239***	−0.055***	0.312
Other	5.296***	−0.642***	0.529	6.903***	−0.687***	−0.120***	−0.259***	0.550	6.847***	−0.686***	−0.120***	−0.257***	0.005	0.550

Table 4.11 compiles the findings regarding the relative performance of the MA strategy versus the BH strategy as well as the trading intensity involved. Overall, the picture that emerges is that even when the strategy works well, skipping one day before trading can negatively impact on performance. For example, the highest performance improvement is for the loser momentum portfolio which is 11.4% per year which is a reduction compared to the baseline implementation. Furthermore, when the strategy does not work so well, skipping a day before trading leads to many portfolios having a negative improvement though the magnitude of the relative loss rarely exceeds 5% per year.

Finally, Table 4.12 presents the regression estimates for the two most commonly used market timing regressions. The coefficients of primary interest here are β_{m^2} and γ_m. In the vast majority of the cases the estimates for β_{m^2} are positive and statistically significant, indicating that the MA strategy has substantial market timing power. However, when we consider the estimates for γ_m we see the source of the reduced performance when we notice that a few of these coefficients are negative and statistically significant. This indicates for those portfolios the MA strategy manages to be on the wrong side of the market on average. The notable exception to this case are the portfolios sorted by volatility for which both sets of parameter estimates are positive and statistically significant.

Figures 4.15 through 4.21 present the scatter plots of the daily returns of the MA strategy versus the daily returns of the BH strategy when we skip one day before we trade for the size-sorted deciles, book-to-market sorted deciles, momentum-sorted deciles, short-term reversal deciles, long-term reversal deciles, volatility-sorted deciles, and industry portfolios, respectively.

4.4 Zero Cash Rate

In this case, the returns of the moving average switching strategy can be expressed as follows:

$$\tilde{R}_{jt,L} = \begin{cases} R_{jt}, & \text{if } P_{jt-1} > A_{jt-1,L}, \\ 0, & \text{otherwise}, \end{cases} \tag{4.5}$$

Table 4.11 Trading intensity with skipping a day

This table reports the results for the improvement delivered by the MA switching strategy over the buy-and-hold strategy one day before trading is skipped, the trading frequency, as well as the BETC using ten decile portfolios sorted by various characteristics. The sample period covers January 4, 1960 until December 31, 2013 with value-weighted portfolio returns. $\Delta\mu$ is the annualized improvement in the average in-sample monthly return, $\Delta\sigma$ is the annualized improvement in the return standard deviation, p_A is the proportion of months during which there is a hold signal, NT is the number of transactions (buy or sell) over the entire sample period, BETC is the break-even one-sided transaction cost in percent, p_1 is the proportion of months during which a buy signal was followed by a positive return of the underlying portfolio and p_2 is the proportion of months during which a buy signal was followed by a portfolio return in excess of the risk-free rate. The length of the MA window is 20 days.

Portfolio	$\Delta\mu$	$\Delta\sigma$	p_A	**NT**	**BETC**	p_1	p_2
			Panel A: Size-sorted portfolios				
Low	10.54	5.08	0.60	931	0.61	75.59	57.78
2	6.95	6.12	0.61	1071	0.35	73.29	56.44
3	4.67	6.14	0.61	1089	0.23	72.55	56.48
4	4.37	5.97	0.62	1147	0.21	72.28	56.08
5	2.62	5.91	0.62	1139	0.12	71.57	56.05
6	1.96	5.53	0.62	1165	0.09	71.73	55.84
7	0.90	5.67	0.62	1145	0.04	71.30	55.50
8	0.36	5.64	0.62	1193	0.02	70.39	54.99
9	−1.52	5.71	0.62	1313	−0.06	70.11	54.43
High	−3.97	5.75	0.61	1501	−0.14	68.61	52.85
			Panel B: Book-to-market sorted portfolios				
Low	−0.90	6.56	0.59	1409	−0.03	70.24	52.84
2	−2.05	5.82	0.61	1339	−0.08	69.59	53.19
3	−2.04	5.44	0.61	1345	−0.08	69.33	53.07
4	−0.84	5.68	0.61	1315	−0.03	69.84	53.69
5	−1.58	5.62	0.61	1407	−0.06	68.86	53.15
6	−1.99	5.42	0.62	1259	−0.09	69.07	53.77
7	−2.00	5.14	0.62	1301	−0.08	69.34	53.95
8	−1.91	5.48	0.63	1315	−0.08	69.25	54.23
9	−0.36	5.56	0.63	1309	−0.01	69.46	54.11
High	−0.31	6.08	0.62	1281	−0.01	69.53	53.97

continued

Table 4.11 Continued

Portfolio	$\Delta\mu$	$\Delta\sigma$	p_A	NT	BETC	p_1	p_2
			Panel C: Momentum-sorted portfolios				
Low	11.41	9.27	0.52	1197	0.52	72.96	50.78
2	4.11	7.43	0.56	1295	0.17	71.13	51.25
3	1.21	5.98	0.58	1315	0.05	70.68	52.40
4	0.40	5.85	0.59	1359	0.02	70.12	52.55
5	−0.97	5.46	0.60	1361	−0.04	69.94	52.98
6	−1.67	5.32	0.62	1401	−0.06	69.51	53.48
7	−2.12	5.40	0.62	1355	−0.08	69.26	53.58
8	−3.66	5.26	0.63	1389	−0.14	69.18	54.65
9	−3.63	5.92	0.62	1435	−0.14	69.14	54.20
High	−4.61	7.48	0.64	1307	−0.19	70.07	55.74
			Panel D: Short-term reversal sorted portfolios				
Low	−12.59	10.62	0.72	1241	−0.55	68.18	57.69
2	−4.91	8.91	0.64	1277	−0.21	69.16	54.68
3	−2.64	7.45	0.62	1345	−0.11	69.27	53.64
4	−1.37	6.59	0.61	1299	−0.06	69.83	53.37
5	−1.33	5.99	0.61	1355	−0.05	69.32	53.45
6	−1.53	5.48	0.62	1315	−0.06	69.38	53.56
7	−0.16	5.22	0.60	1413	−0.01	69.95	53.13
8	0.10	4.80	0.58	1401	0.00	70.12	52.97
9	−1.81	4.83	0.57	1409	−0.07	70.68	53.08
High	6.68	6.17	0.47	1475	0.25	73.73	50.48
			Panel E: Long-term reversal sorted portfolios				
Low	2.22	6.98	0.58	1295	0.09	71.08	53.33
2	−1.50	5.95	0.61	1309	−0.06	69.70	53.92
3	−1.72	5.31	0.62	1429	−0.07	69.14	53.53
4	−2.45	5.37	0.62	1389	−0.10	68.93	53.57
5	−1.54	5.30	0.62	1363	−0.06	69.29	53.53
6	−2.31	4.95	0.63	1311	−0.10	68.57	53.51
7	−2.69	5.04	0.62	1407	−0.10	69.12	53.95
8	−2.01	5.65	0.61	1397	−0.08	69.35	53.05
9	−2.00	6.21	0.60	1425	−0.08	69.61	52.84
High	−1.37	7.86	0.59	1375	−0.05	70.35	53.36

continued

Table 4.11 Continued

Portfolio	$\Delta\mu$	$\Delta\sigma$	p_A	NT	BETC	p_1	p_2
			Panel F: Volatility-sorted portfolios				
High	8.78	6.28	0.65	935	0.51	73.23	57.91
2	8.42	7.06	0.61	1097	0.42	73.29	56.13
3	7.07	6.74	0.61	1041	0.37	73.12	56.20
4	4.83	6.24	0.62	1053	0.25	72.92	56.70
5	3.98	5.74	0.63	1025	0.21	72.52	57.10
6	3.25	5.35	0.64	1047	0.17	71.85	57.05
7	2.42	4.77	0.65	1091	0.12	72.22	57.61
8	3.11	4.21	0.66	989	0.17	71.69	57.72
9	3.02	3.61	0.67	927	0.18	72.02	58.60
Low	3.36	2.70	0.68	831	0.22	73.54	59.99
			Panel G: Industry-sorted portfolios				
NoDur	0.12	4.63	0.63	1299	0.01	70.23	54.62
Durbl	−1.40	6.93	0.57	1391	−0.05	69.80	51.15
Manuf	−1.22	5.80	0.61	1315	−0.05	69.37	53.22
Enrgy	−5.41	6.41	0.59	1431	−0.21	69.06	52.46
HiTec	−0.20	8.26	0.57	1407	−0.01	71.05	52.39
Telcm	−3.02	5.86	0.58	1461	−0.11	69.51	51.28
Shops	−0.34	5.73	0.60	1313	−0.01	70.35	53.47
Hlth	−2.64	5.68	0.60	1331	−0.11	70.14	53.13
Utils	1.83	4.31	0.61	1213	0.08	70.95	53.87
Other	1.42	6.47	0.61	1273	0.06	70.51	53.94

in the absence of any transaction costs imposed on the switches. The alternative specification for the case of positive one-way transaction cost of τ leads to the following four cases in the post-transaction cost returns:

$$\tilde{R}_{jt,L} = \begin{cases} R_{jt}, & \text{if } P_{jt-1} > A_{jt-1,L} \text{ and } P_{jt-2} > A_{jt-2,L}, \\ R_{jt} - \tau, & \text{if } P_{jt-1} > A_{jt-1,L} \text{ and } P_{jt-2} < A_{jt-2,L}, \\ 0, & \text{if } P_{jt-1} < A_{jt-1,L} \text{ and } P_{jt-2} < A_{jt-2,L}, \\ -\tau, & \text{if } P_{jt-1} < A_{jt-1,L} \text{ and } P_{jt-2} > A_{jt-2,L}. \end{cases} \tag{4.6}$$

Table 4.13 presents the summary statistics for the performance of the MA, BH, and MAP returns when we use cash instead of the risk-free asset. It is instructive to compare these to the results for the baseline implementation reported in Table 2.1. As expected,

Table 4.12 Market timing with skipping a day

This table reports alphas, betas, and adjusted R^2 of the market-timing regressions of the MAP excess returns on the market factor using portfolios sorted by various characteristics when a day is skipped before trading. The TM panel reports the results using the Treynor and Mazuy (1966) quadratic regression with the squared market factor (β_{m^2}) while the HM panel reports the results using the Merton and Henriksson (1981) regression with option-like returns on the market (γ_m). The sample period covers January 4, 1960 until December 31, 2013 with value-weighted portfolio returns. The length of the moving average window is 20 days. Newey and West (1987) standard errors with five lags are used in reporting statistical significance of a two-sided null hypothesis at the 1%, 5%, and 10% level given by a***, a**, and a*, respectively.

Portfolio	α	β_m	β_{m^2}	$\overline{R}^2$	α	β_m	γ_m	$\overline{R}^2$
			Panel A: Size-sorted portfolios					
Low	10.039***	−0.423***	0.013***	0.405	2.119	−0.359***	0.133***	0.404
2	8.273***	−0.555***	0.008***	0.454	3.000	−0.513***	0.089***	0.454
3	6.620***	−0.573***	0.006***	0.483	2.989*	−0.544***	0.063***	0.483
4	5.851***	−0.567***	0.008***	0.494	2.234	−0.535***	0.068***	0.493
5	4.699***	−0.574***	0.006***	0.511	2.711	−0.555***	0.041***	0.511
6	3.342***	−0.546***	0.008***	0.526	0.828	−0.522***	0.054***	0.524
7	2.588**	−0.564***	0.007***	0.541	1.681	−0.551***	0.032**	0.540
8	1.785	−0.573***	0.008***	0.553	0.039	−0.553***	0.046***	0.551
9	−0.172	−0.585***	0.009***	0.583	0.409	−0.579***	0.020*	0.581
High	−1.888*	−0.600***	0.006***	0.580	1.649	−0.615***	−0.024**	0.579

Panel B: Book-to-market sorted portfolios

Low	2.797**	−0.661***	0.001	0.564	6.137***	−0.681***	−0.037***	0.564
2	−0.057	−0.596***	0.007***	0.565	2.265	−0.604***	−0.008	0.564
3	0.066	−0.554***	0.005***	0.539	2.983**	−0.567***	−0.020	0.539
4	−0.751	−0.575***	0.014***	0.558	−1.267	−0.556***	0.048***	0.553
5	0.629	−0.556***	0.005***	0.522	2.100	−0.561***	−0.004	0.522
6	−0.527	−0.537***	0.007***	0.523	0.654	−0.536***	0.008	0.521
7	−0.857	−0.507***	0.008***	0.498	0.219	−0.506***	0.010	0.496
8	−1.084	−0.526***	0.010***	0.476	−0.794	−0.518***	0.025*	0.473
9	−0.210	−0.537***	0.013***	0.488	−1.910	−0.513***	0.059***	0.484
High	0.348	−0.578***	0.012***	0.461	−0.593	−0.560***	0.046***	0.458

continued

Table 4.12 Continued

Portfolio	α	β_m	β_{m^2}	$\bar{R}^2$	α	β_m	γ_m	$\bar{R}^2$
			Panel C: Momentum-sorted portfolios					
Low	16.628***	−0.844***	−0.001	0.456	17.578***	−0.850***	−0.013	0.456
2	8.698***	−0.716***	−0.001	0.498	10.785***	−0.730***	−0.029*	0.498
3	5.747***	−0.609***	−0.004***	0.509	9.231***	−0.635***	−0.053***	0.509
4	2.816**	−0.592***	0.005***	0.531	4.076**	−0.595***	−0.001	0.530
5	2.572**	−0.559***	−0.001	0.523	5.999***	−0.581***	−0.044***	0.523
6	−0.386	−0.535***	0.008***	0.519	1.958	−0.541***	−0.005	0.517
7	−0.474	−0.543***	0.007***	0.532	1.342	−0.547***	−0.002	0.531
8	−2.973**	−0.530***	0.010***	0.512	−1.785	−0.527***	0.016	0.509
9	−2.371*	−0.586***	0.009***	0.526	−2.554	−0.575***	0.030**	0.524
High	−2.586	−0.690***	0.009***	0.466	−0.864	−0.691***	0.005	0.465

Panel D: Short-term reversal sorted portfolios

Low	−4.981**	−0.844***	−0.010***	0.394	8.301***	−0.939***	−0.192***	0.395
2	−0.214	−0.768***	−0.000	0.500	5.152**	−0.802***	−0.066***	0.500
3	1.333	−0.672***	0.000	0.512	5.905***	−0.701***	−0.054***	0.513
4	1.511	−0.619***	0.004***	0.527	6.842***	−0.649***	−0.054***	0.528
5	0.747	−0.583***	0.006***	0.539	3.638**	−0.595***	−0.017	0.538
6	−0.188	−0.556***	0.008***	0.538	1.445	−0.558***	0.005	0.536
7	0.741	−0.553***	0.010***	0.561	0.788	−0.542***	0.029**	0.558
8	1.831*	−0.532***	0.006***	0.542	0.524	−0.517***	0.034***	0.541
9	−0.160	−0.541***	0.007***	0.525	−1.633	−0.525***	0.038***	0.523
High	8.385***	−0.648***	0.009***	0.518	0.712	−0.589***	0.120***	0.519

continued

Table 4.12 Continued

Portfolio	α	β_m	β_{m^2}	$\overline{R}^2$	α	β_m	γ_m	$\overline{R}^2$
			Panel E: Long-term reversal sorted portfolios					
Low	3.892**	−0.675***	0.010***	0.483	3.582*	−0.662***	0.033*	0.482
2	−1.063	−0.590***	0.013***	0.512	−0.397	−0.580***	0.030**	0.508
3	−0.748	−0.538***	0.009***	0.503	−0.030	−0.532***	0.019	0.501
4	−2.570**	−0.537***	0.014***	0.515	−1.818	−0.527***	0.032**	0.510
5	−1.782	−0.532***	0.014***	0.519	−0.520	−0.525***	0.027**	0.512
6	−0.877	−0.501***	0.007***	0.506	0.610	−0.504***	0.002	0.505
7	−0.475	−0.516***	0.004***	0.512	1.386	−0.524***	−0.011	0.511
8	−0.183	−0.564***	0.007***	0.537	0.507	−0.561***	0.011	0.536
9	1.841	−0.612***	−0.001	0.528	4.608***	−0.630***	−0.035***	0.528
High	4.430***	−0.765***	−0.005***	0.544	7.817***	−0.792***	−0.056***	0.544

Panel F: Volatility-sorted portfolios								
High	9.448***	−0.542***	0.011***	0.342	3.224	−0.491***	0.107***	0.341
2	9.864***	−0.645***	0.010***	0.461	4.416**	−0.600***	0.096***	0.460
3	8.817***	−0.626***	0.008***	0.480	5.707***	−0.597***	0.063***	0.479
4	6.506***	−0.583***	0.008***	0.485	4.433**	−0.562***	0.048***	0.484
5	5.291***	−0.537***	0.008***	0.489	2.556	−0.512***	0.057***	0.488
6	3.939***	−0.503***	0.010***	0.499	0.868	−0.473***	0.066***	0.497
7	3.113***	−0.444***	0.008***	0.491	1.759	−0.427***	0.041***	0.489
8	3.072***	−0.380***	0.010***	0.475	0.681	−0.355***	0.058***	0.471
9	2.708***	−0.299***	0.009***	0.418	0.221	−0.274***	0.056***	0.414
Low	2.138***	−0.169***	0.009***	0.256	−0.898	−0.140***	0.064***	0.249
Panel G: Industry-sorted portfolios								
NoDur	−0.052	−0.433***	0.012***	0.447	−1.784	−0.410***	0.055***	0.442
Durbl	0.624	−0.679***	0.009***	0.473	1.997	−0.678***	0.009	0.472
Manuf	−0.392	−0.587***	0.011***	0.547	0.658	−0.582***	0.021	0.544
Enrgy	−3.398**	−0.539***	0.005***	0.328	−1.353	−0.546***	−0.010	0.328
HiTec	4.381***	−0.783***	0.001	0.488	9.723***	−0.816***	−0.063***	0.489
Telcm	−0.000	−0.530***	0.001	0.401	4.576**	−0.558***	−0.053***	0.401
Shops	2.087	−0.551***	0.004***	0.469	4.559***	−0.562***	−0.019	0.469
Hlth	−1.455	−0.514***	0.008***	0.407	0.150	−0.515***	0.004	0.405
Utils	7.272***	−0.335***	−0.014***	0.289	7.475***	−0.351***	−0.044***	0.280
Other	3.813***	−0.639***	0.006***	0.529	3.866**	−0.633***	0.017	0.529

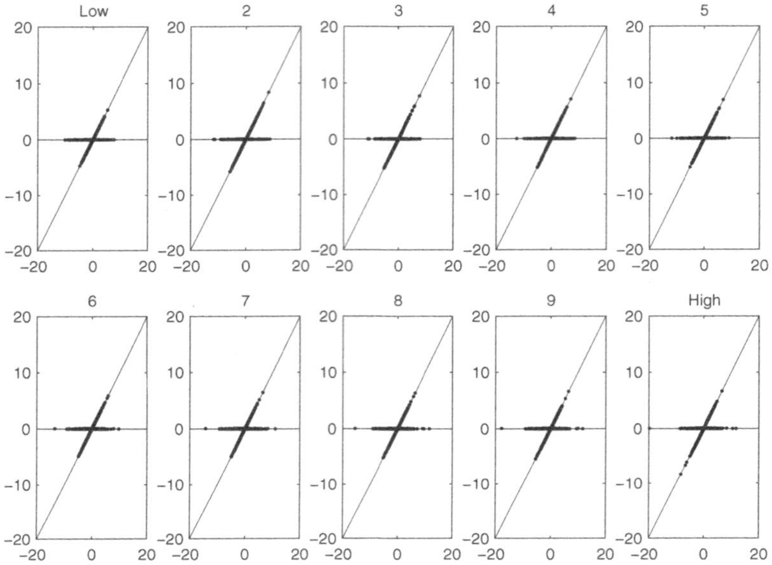

Figure 4.15 Scatter plot of BH versus MA returns with skipping a day: Size reversal decile portfolios.

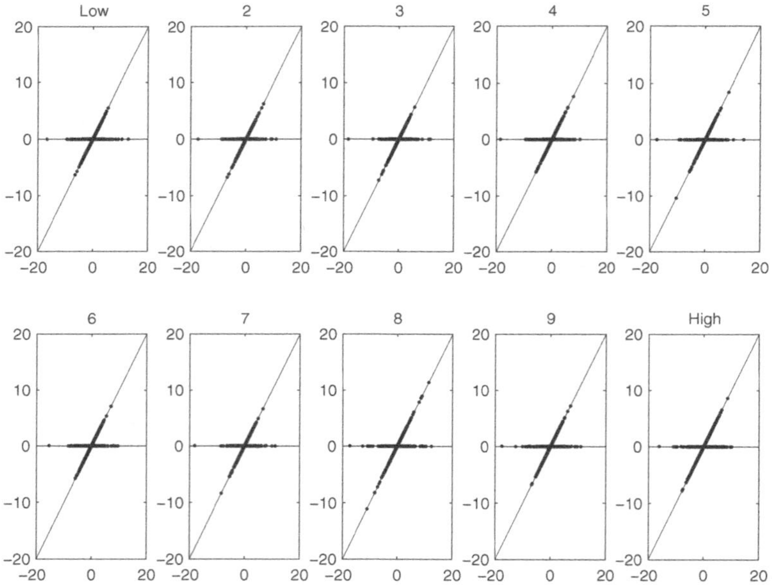

Figure 4.16 Scatter plot of BH versus MA returns with skipping a day: Book-to-market reversal decile portfolios.

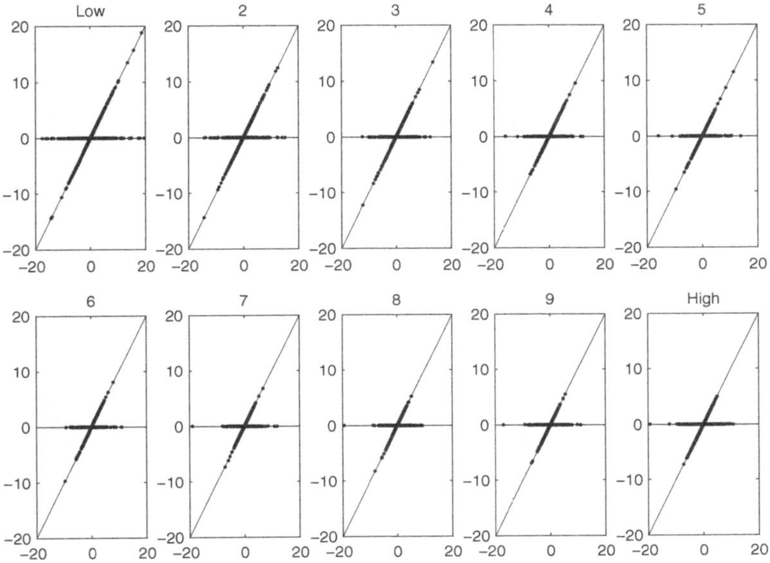

Figure 4.17 Scatter plot of BH versus MA returns with skipping a day: Momentum reversal decile portfolios.

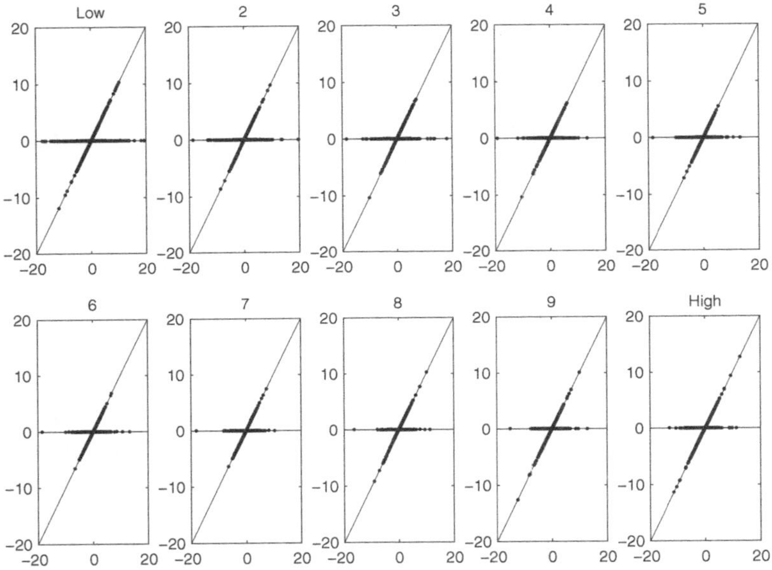

Figure 4.18 Scatter plot of BH versus MA returns with skipping a day: Short-term reversal decile portfolios.

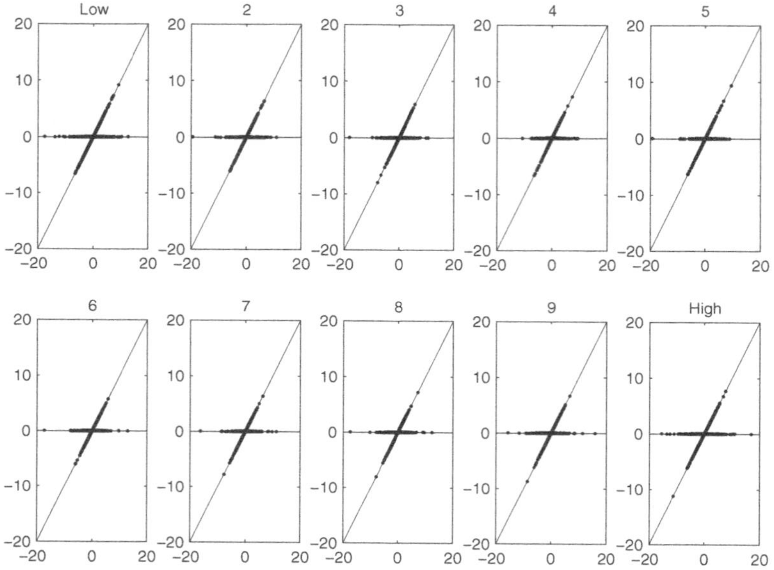

Figure 4.19 Scatter plot of BH versus MA returns with skipping a day: Long-term reversal decile portfolios.

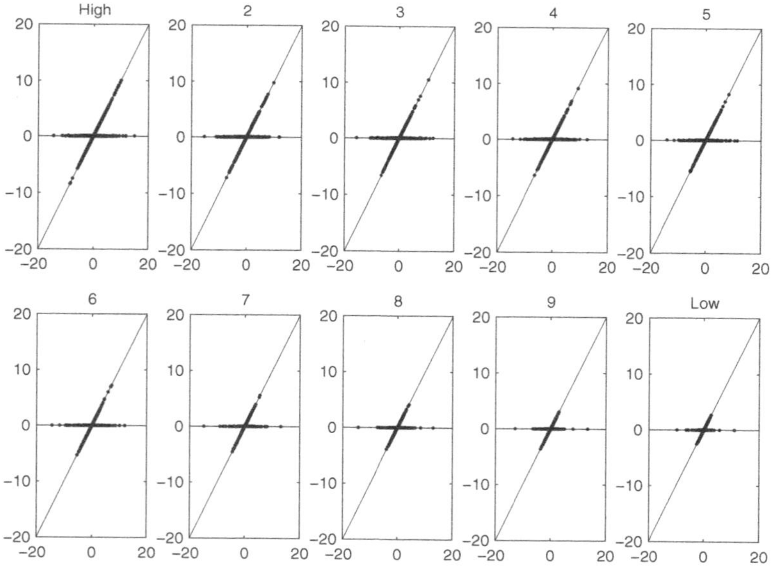

Figure 4.20 Scatter plot of BH versus MA returns with skipping a day: Volatility decile portfolios.

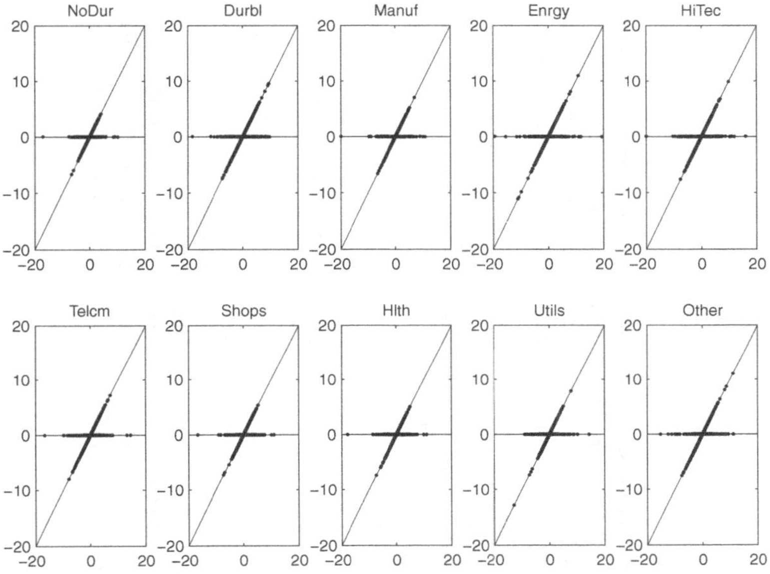

Figure 4.21 Scatter plot of BH versus MA returns with skipping a day: Industry portfolios.

foregoing the risk-free interest rate reduces the average returns of the MA strategy relative to the BH strategy by between 1% and 2% per year on average. This leads to a reduced outperformance of the MA strategy relative to the BH strategy in more cases than it does using the baseline implementation.

Table 4.14 reports the parameter estimates for the CAPM model, the Fama-French three-factor model, and the Carhart four-factor model. Once again, the number and magnitude of positive and statistically significant abnormal returns are reduced when compared to the baseline case investigated in the previous section and chapters. The level of annualized αs can be as high as 18% per year for the loser momentum decile when using the CAPM model. This leads to the conclusion that foregoing to risk-free rate results in reduced performance of the MA strategy compared to the BH strategy.

Table 4.15 compiles the findings regarding the relative performance of the MA strategy versus the BH strategy as well as the trading intensity involved. Overall, the picture that emerges is that even when the strategy works well, skipping one day before trading can negatively impact on performance. For example, the highest

Table 4.13 Portfolio performance with zero cash rate

This table reports summary statistics for the respective buy-and-hold (BH) portfolio returns, the MA switching strategy portfolio returns with a zero cash rate, and the excess return of MA over BH (MAP) using sets of ten portfolios sorted by various characteristics. The sample period covers January 4, 1960 until December 31, 2013 with value-weighted portfolio returns. μ is the annualized average return, σ is the annualized standard deviation of returns, s is the annualized skewness, k is the annualized kurtosis, and SR is the annualized Sharpe ratio. The length of the MA window is 20 days.

	BH Portfolios					**MA Portfolios**					**MAP Portfolios**				
Portfolio	μ	σ	s	k	SR	μ	σ	s	k	SR	μ	σ	s	k	SR
						Panel A: Size-sorted portfolios									
Low	11.88	13.58	−0.90	14.81	0.52	23.79	8.36	−0.24	12.56	2.27	11.91	10.60	1.49	34.15	1.12
2	11.86	16.72	−0.46	13.85	0.42	20.04	10.54	−0.57	22.19	1.45	8.19	12.92	0.47	28.76	0.63
3	12.96	16.60	−0.49	12.02	0.49	19.68	10.47	−0.12	13.29	1.42	6.72	12.83	0.76	27.37	0.52
4	12.18	16.33	−0.48	12.48	0.45	18.23	10.33	−0.44	15.99	1.30	6.05	12.62	0.59	27.61	0.48
5	12.68	16.23	−0.47	11.96	0.49	17.23	10.32	−0.54	15.85	1.21	4.54	12.50	0.51	26.43	0.36
6	12.26	15.33	−0.53	13.43	0.49	16.66	9.74	−0.47	14.54	1.22	4.40	11.81	0.68	31.11	0.37
7	12.29	15.47	−0.55	15.54	0.49	16.37	9.82	−0.42	15.53	1.18	4.08	11.94	0.76	36.49	0.34
8	12.01	15.70	−0.51	16.62	0.46	14.66	10.00	−0.37	14.90	0.99	2.65	12.09	0.70	40.02	0.22
9	11.57	15.46	−0.59	20.90	0.44	11.79	9.75	−0.14	12.02	0.72	0.22	12.00	1.00	52.11	0.02
High	10.23	16.00	−0.50	21.64	0.34	7.07	10.15	−0.16	13.65	0.23	−3.16	12.38	0.84	54.11	−0.26

Panel B: Book-to-market sorted portfolios															
Low	9.49	17.53	−0.18	13.47	0.27	10.37	10.94	−0.01	10.08	0.51	0.88	13.69	0.23	32.00	0.06
2	11.03	16.07	−0.39	16.37	0.39	10.99	10.25	−0.14	12.76	0.61	−0.04	12.38	0.60	40.34	−0.00
3	11.15	15.42	−0.52	20.70	0.41	11.43	9.94	−0.04	11.54	0.67	0.29	11.79	0.95	54.48	0.02
4	10.95	15.80	−0.71	20.94	0.39	11.61	10.22	−0.19	16.43	0.67	0.66	12.05	1.31	53.09	0.05
5	11.13	15.62	−0.45	23.02	0.41	9.50	10.10	−0.42	19.49	0.47	−1.63	11.91	0.58	57.72	−0.14
6	11.89	15.14	−0.45	16.96	0.47	11.52	9.82	−0.29	15.45	0.69	−0.36	11.53	0.63	42.17	−0.03
7	12.55	14.89	−0.64	22.73	0.52	12.07	9.76	−0.32	14.82	0.75	−0.47	11.25	1.07	61.04	−0.04
8	12.98	15.82	−0.60	30.16	0.52	12.28	10.27	−0.47	33.33	0.73	−0.71	12.03	0.88	72.10	−0.06
9	13.90	15.99	−0.68	20.87	0.57	13.69	10.49	−0.49	17.87	0.85	−0.21	12.07	1.04	53.74	−0.02
High	15.01	17.78	−0.46	15.88	0.58	16.76	11.74	−0.29	17.42	1.02	1.75	13.34	0.66	39.37	0.13
Panel C: Momentum-sorted portfolios															
Low	2.59	24.93	0.41	25.70	−0.09	16.37	15.87	1.11	58.82	0.73	13.78	19.19	−0.34	45.73	0.72
2	8.03	20.35	0.15	22.17	0.16	14.42	13.27	−0.03	40.56	0.73	6.39	15.40	−0.49	45.30	0.41
3	10.29	17.59	0.04	17.98	0.31	13.79	11.69	0.36	31.54	0.77	3.50	13.13	0.00	38.06	0.27
4	10.22	16.61	−0.15	18.81	0.33	11.82	10.86	0.44	19.83	0.65	1.60	12.56	0.48	46.26	0.13
5	10.09	15.89	0.01	21.65	0.33	9.93	10.48	0.20	25.95	0.49	−0.17	11.95	−0.04	52.44	−0.01
6	10.99	15.35	−0.58	26.71	0.40	10.38	10.03	−0.31	19.29	0.56	−0.61	11.62	0.96	70.31	−0.05
7	10.92	15.19	−0.55	22.88	0.40	9.99	9.79	−0.20	12.52	0.53	−0.93	11.62	0.93	60.32	−0.08
8	13.57	15.49	−0.73	21.21	0.57	11.44	10.13	−0.34	11.30	0.66	−2.13	11.73	1.26	57.86	−0.18
9	12.81	16.47	−0.54	15.10	0.49	11.55	10.41	−0.32	10.04	0.65	−1.26	12.76	0.79	37.23	−0.10
High	17.85	20.37	−0.50	12.74	0.64	18.01	12.65	−0.56	9.54	1.05	0.16	15.97	0.55	29.81	0.01

continued

Table 4.13 Continued

Portfolio	BH Portfolios					MA Portfolios					MAP Portfolios				
	μ	σ	s	k	SR	μ	σ	s	k	SR	μ	σ	s	k	SR
Panel D: Short-term reversal sorted portfolios															
Low	41.02	25.55	0.87	29.76	1.42	33.81	15.13	0.14	22.97	1.92	−7.21	20.64	−1.95	64.00	−0.35
2	18.83	20.36	0.33	27.47	0.69	17.13	11.69	0.11	23.35	1.06	−1.69	16.68	−0.77	55.57	−0.10
3	14.32	17.97	0.06	26.89	0.53	13.51	10.58	−0.32	23.82	0.83	−0.82	14.53	−0.41	56.28	−0.06
4	11.57	16.65	−0.38	23.86	0.41	12.52	10.09	0.01	19.14	0.77	0.95	13.24	0.60	53.07	0.07
5	11.73	15.81	−0.47	22.77	0.44	12.04	9.72	0.06	13.28	0.75	0.31	12.46	0.81	53.78	0.02
6	12.78	15.57	−0.45	22.73	0.51	12.90	10.09	−0.02	12.46	0.80	0.11	11.86	0.80	60.78	0.01
7	9.63	15.45	−0.55	19.99	0.31	9.57	10.23	−0.25	15.05	0.47	−0.06	11.58	0.97	53.93	−0.00
8	7.93	15.66	−0.42	16.68	0.20	8.96	10.79	−0.21	19.68	0.39	1.04	11.34	0.79	44.32	0.09
9	6.01	16.52	−0.44	15.75	0.07	6.32	11.70	−0.69	23.27	0.13	0.31	11.66	0.46	39.80	0.03
High	−9.59	19.08	−0.55	11.88	−0.75	−1.41	12.80	−1.08	25.77	−0.48	8.17	14.16	0.66	22.13	0.58
Panel E: Long-term reversal sorted portfolios															
Low	14.73	20.02	−0.38	14.61	0.50	19.27	13.06	0.24	12.85	1.11	4.55	15.15	0.83	37.12	0.30
2	13.72	17.10	−0.61	18.75	0.52	14.05	11.15	−0.32	14.54	0.83	0.33	12.97	0.99	48.47	0.03
3	13.09	15.94	−0.58	18.32	0.52	12.15	10.62	−0.15	12.17	0.69	−0.93	11.89	1.08	51.10	−0.08
4	12.96	15.66	−0.79	25.23	0.52	11.14	10.36	−0.22	13.61	0.61	−1.82	11.75	1.53	70.99	−0.15
5	11.63	15.50	−0.71	22.89	0.44	10.85	10.26	−0.08	17.67	0.59	−0.77	11.62	1.44	61.32	−0.07
6	12.91	14.73	−0.36	21.45	0.55	11.40	9.75	−0.17	12.59	0.68	−1.51	11.04	0.53	60.02	−0.14
7	12.10	15.03	−0.39	17.81	0.49	10.78	9.97	−0.13	12.91	0.60	−1.32	11.25	0.64	48.54	−0.12
8	11.03	15.60	−0.46	19.22	0.40	10.28	9.95	−0.09	13.30	0.55	−0.75	12.02	0.80	48.15	−0.06
9	9.75	16.87	−0.19	18.39	0.29	10.07	10.57	−0.15	13.84	0.50	0.32	13.14	0.17	44.08	0.02
High	10.66	20.33	−0.16	14.88	0.29	12.52	12.41	−0.30	13.98	0.62	1.87	16.10	0.06	32.89	0.12

						Panel F: Volatility-sorted portfolios									
High	40.10	20.27	0.10	14.40	1.74	51.20	14.08	1.36	18.48	3.30	11.09	14.42	0.44	38.43	0.77
2	17.80	19.30	−0.31	14.52	0.67	29.42	12.25	0.59	17.19	2.01	11.62	14.82	0.75	33.20	0.78
3	15.22	18.25	−0.38	16.39	0.57	25.35	11.51	0.08	20.51	1.79	10.12	14.10	0.61	36.62	0.72
4	15.74	16.89	−0.47	18.77	0.65	23.05	10.72	−0.15	23.68	1.70	7.31	13.00	0.70	42.16	0.56
5	15.68	15.52	−0.55	20.47	0.70	22.13	9.87	−0.26	25.24	1.76	6.46	11.92	0.78	46.43	0.54
6	14.82	14.38	−0.67	24.45	0.70	19.69	9.17	−0.54	28.36	1.63	4.87	11.05	0.89	56.32	0.44
7	14.40	12.78	−0.66	28.32	0.75	19.00	8.08	−0.35	20.40	1.76	4.60	9.86	0.93	70.12	0.47
8	13.68	11.15	−0.78	36.32	0.80	18.49	6.93	0.11	11.61	1.98	4.81	8.70	1.36	92.78	0.55
9	12.53	9.27	−0.58	54.72	0.84	17.25	5.63	0.32	11.22	2.21	4.72	7.32	0.96	136.30	0.65
Low	10.81	6.86	−0.47	71.59	0.88	15.27	4.14	0.71	14.99	2.54	4.46	5.43	0.82	177.24	0.82
						Panel G: Industry-sorted portfolios									
NoDur	12.80	13.63	−0.62	22.17	0.59	13.83	9.05	−0.12	12.13	1.00	1.02	10.18	1.17	63.09	0.10
Durbl	10.57	20.63	−0.23	11.77	0.28	11.02	13.75	0.16	15.06	0.45	0.45	15.38	0.54	28.39	0.03
Manuf	11.07	16.28	−0.68	21.25	0.39	12.88	10.47	−0.23	14.50	0.77	1.80	12.46	1.20	54.42	0.14
Enrgy	13.49	19.99	−0.20	19.86	0.44	10.55	13.45	−0.25	14.31	0.43	−2.93	14.79	0.15	56.37	−0.20
HiTec	11.37	22.42	0.02	12.00	0.29	13.35	14.14	0.19	10.65	0.61	1.97	17.38	−0.08	28.50	0.11
Telcm	10.94	17.36	−0.03	17.12	0.36	8.98	11.49	0.13	13.85	0.37	−1.97	13.02	0.02	45.75	−0.15
Shops	12.23	16.54	−0.28	14.21	0.45	13.91	10.85	0.06	12.60	0.84	1.68	12.48	0.50	36.48	0.13
Hlth	12.52	16.83	−0.40	14.37	0.46	12.61	11.10	−0.10	11.24	0.71	0.10	12.65	0.70	38.21	0.01
Utils	10.18	12.93	−0.03	28.88	0.42	12.10	8.34	−0.20	16.38	0.88	1.93	9.87	−0.27	76.72	0.20
Other	11.15	17.73	−0.26	18.07	0.36	13.69	11.20	−0.11	29.02	0.80	2.54	13.74	0.34	37.22	0.19

Table 4.14 Factor loadings with zero cash rate

This table reports alphas, betas, and adjusted R^2 of the regressions of the MAP excess returns on the market factor, the Fama-French three-factors, and the Carhart four-factors using portfolios sorted by various characteristics. The alphas are annualized and in percent. The sample period covers January 4, 1960 until December 31, 2013 with daily value-weighted portfolio returns. The length of the MA window is 20 days. Newey and West (1987) standard errors with five lags are used in reporting statistical significance of a two-sided null hypothesis at the 1%, 5%, and 10% level given by a***, a**, and a*, respectively.

	CAPM			Fama-French					Carhart					
Portfolio	α	β_m	$\bar{R}^2$	α	β_m	β_s	β_b	$\bar{R}^2$	α	β_m	β_s	β_b	β_u	$\bar{R}^2$
							Panel A: Size-sorted portfolios							
Low	14.531***	−0.435***	0.405	16.195***	−0.491***	−0.522***	−0.133***	0.554	17.122***	−0.503***	−0.520***	−0.164***	−0.087***	0.561
2	11.572***	−0.561***	0.453	13.465***	−0.625***	−0.600***	−0.149***	0.585	14.421***	−0.638***	−0.597***	−0.181***	−0.089***	0.590
3	10.191***	−0.575***	0.483	11.885***	−0.633***	−0.555***	−0.127***	0.596	12.765***	−0.644***	−0.553***	−0.157***	−0.082***	0.601
4	9.497***	−0.572***	0.494	10.864***	−0.620***	−0.500***	−0.085***	0.589	11.868***	−0.633***	−0.498***	−0.118***	−0.094***	0.595
5	8.018***	−0.576***	0.511	9.105***	−0.615***	−0.449***	−0.050***	0.588	9.913***	−0.625***	−0.447***	−0.077***	−0.075***	0.592
6	7.741***	−0.553***	0.528	8.427***	−0.579***	−0.315***	−0.021**	0.571	9.287***	−0.590***	−0.313***	−0.050***	−0.080***	0.576
7	7.493***	−0.565***	0.540	8.176***	−0.589***	−0.257***	−0.040***	0.568	9.092***	−0.601***	−0.254***	−0.071***	−0.086***	0.573
8	6.171***	−0.583***	0.559	6.780***	−0.603***	−0.179***	−0.053***	0.573	7.644***	−0.614***	−0.177***	−0.082***	−0.081***	0.578
9	3.787***	−0.590***	0.583	4.244***	−0.604***	−0.061***	−0.064***	0.585	5.095***	−0.615***	−0.059***	−0.093***	−0.079***	0.590
High	0.536	−0.612***	0.588	−0.053	−0.593***	0.147***	0.060***	0.597	0.459	−0.599***	0.148***	0.043***	−0.048***	0.599

Panel B: Book-to-market sorted portfolios														
Low	4.882***	−0.662***	0.563	3.094***	−0.614***	0.041***	0.320***	0.591	3.543***	−0.620***	0.042***	0.305***	−0.042***	0.592
2	3.574***	−0.599***	0.564	2.972***	−0.583***	0.025***	0.104***	0.568	3.739***	−0.593***	0.027***	0.078***	−0.072***	0.572
3	3.675***	−0.561***	0.546	3.440***	−0.555***	0.024***	0.036***	0.546	4.504***	−0.568***	0.027***	0.000	−0.099***	0.554
4	4.119***	−0.574***	0.546	4.818***	−0.592***	−0.005	−0.129***	0.552	5.553***	−0.602***	−0.003	−0.153***	−0.069***	0.555
5	1.707	−0.553***	0.519	2.676**	−0.579***	−0.009	−0.178***	0.531	3.278***	−0.587***	−0.008	−0.198***	−0.056***	0.533
6	2.862**	−0.534***	0.517	3.940***	−0.564***	−0.070***	−0.178***	0.531	4.541***	−0.572***	−0.068***	−0.198***	−0.056***	0.533
7	2.622**	−0.513***	0.500	4.016***	−0.550***	−0.031***	−0.250***	0.525	4.888***	−0.562***	−0.029***	−0.279***	−0.081***	0.531
8	2.498**	−0.531***	0.469	4.930***	−0.597***	−0.072***	−0.429***	0.534	5.296***	−0.601***	−0.071***	−0.442***	−0.034***	0.535
9	3.046**	−0.540***	0.482	5.166***	−0.597***	−0.064***	−0.374***	0.531	6.193***	−0.611***	−0.061***	−0.408***	−0.096***	0.538
High	5.264***	−0.582***	0.457	7.897***	−0.655***	−0.184***	−0.429***	0.517	8.869***	−0.668***	−0.182***	−0.461***	−0.091***	0.522
Panel C: Momentum-sorted portfolios														
Low	18.791***	−0.830***	0.451	20.348***	−0.878***	−0.274***	−0.198***	0.467	15.701***	−0.818***	−0.286***	−0.042**	0.434***	0.522
2	10.597***	−0.697***	0.493	11.476***	−0.722***	−0.081***	−0.136***	0.498	8.181***	−0.680***	−0.090***	−0.026**	0.308***	0.542
3	7.148***	−0.604***	0.510	7.850***	−0.623***	−0.001	−0.130***	0.515	5.834***	−0.597***	−0.007	−0.063***	0.188***	0.537
4	5.170***	−0.591***	0.533	5.780***	−0.607***	0.021**	−0.121***	0.538	5.025***	−0.597***	0.019**	−0.096***	0.070***	0.541
5	3.207***	−0.559***	0.527	3.594***	−0.569***	0.001	−0.072***	0.529	3.585***	−0.569***	0.001	−0.072***	0.001	0.529
6	2.651**	−0.540***	0.521	2.927***	−0.547***	0.026***	−0.060***	0.522	3.592***	−0.556***	0.028***	−0.083***	−0.062***	0.525
7	2.371**	−0.546***	0.532	2.650**	−0.553***	0.012	−0.056***	0.533	4.062***	−0.571***	0.015*	−0.103***	−0.132***	0.547
8	1.133	−0.541***	0.512	1.453	−0.549***	0.018**	−0.066***	0.514	3.643***	−0.577***	0.024***	−0.139***	−0.204***	0.546
9	2.334*	−0.596***	0.524	2.535**	−0.603***	−0.078***	−0.011	0.527	5.339***	−0.639***	−0.070***	−0.105***	−0.262***	0.572
High	4.433***	−0.708***	0.473	3.924**	−0.700***	−0.234***	0.175***	0.494	8.308***	−0.757***	−0.222***	0.028**	−0.409***	0.565

continued

Table 4.14 Continued

Portfolio	CAPM α	CAPM β_m	CAPM $\overline{R}^2$	Fama-French α	Fama-French β_m	Fama-French β_s	Fama-French β_b	Fama-French $\overline{R}^2$	Carhart α	Carhart β_m	Carhart β_s	Carhart β_b	Carhart β_u	Carhart $\overline{R}^2$
Panel D: Short-term reversal sorted portfolios														
Low	−2.193	−0.832***	0.391	−1.671	−0.851***	−0.233***	−0.018	0.398	−2.564	−0.839***	−0.235***	0.012	0.083***	0.400
2	2.873*	−0.756***	0.494	3.195**	−0.767***	−0.094***	−0.028**	0.496	3.225**	−0.767***	−0.094***	−0.029**	−0.003	0.496
3	3.216**	−0.668***	0.509	3.505**	−0.677***	−0.049***	−0.037***	0.510	4.174***	−0.686***	−0.048***	−0.059***	−0.062***	0.512
4	4.712***	−0.623***	0.533	4.907***	−0.629***	−0.007	−0.034***	0.534	5.765***	−0.640***	−0.005	−0.063***	−0.080***	0.538
5	3.905***	−0.596***	0.551	4.170***	−0.603***	−0.011	−0.046***	0.552	4.923***	−0.613***	−0.009	−0.071***	−0.070***	0.555
6	3.501***	−0.561***	0.538	3.776***	−0.568***	−0.004	−0.050***	0.539	4.594***	−0.579***	−0.002	−0.077***	−0.076***	0.544
7	3.306***	−0.557***	0.557	3.604***	−0.565***	−0.011	−0.052***	0.558	4.289***	−0.574***	−0.009	−0.075***	−0.064***	0.561
8	4.297***	−0.540***	0.546	4.501***	−0.546***	−0.017*	−0.033***	0.546	5.495***	−0.559***	−0.014*	−0.066***	−0.093***	0.554
9	3.611***	−0.546***	0.529	4.034***	−0.559***	−0.045***	−0.064***	0.531	4.534***	−0.565***	−0.044***	−0.080***	−0.047***	0.533
High	12.160***	−0.661***	0.524	12.634***	−0.678***	−0.182***	−0.026**	0.534	12.659***	−0.678***	−0.182***	−0.027**	−0.002	0.534
Panel E: Long-term reversal sorted portfolios														
Low	8.643***	−0.679***	0.483	10.378***	−0.732***	−0.335***	−0.210***	0.519	11.874***	−0.752***	−0.332***	−0.260***	−0.140***	0.528
2	3.928***	−0.596***	0.509	5.144***	−0.632***	−0.149***	−0.176***	0.525	6.558***	−0.650***	−0.145***	−0.224***	−0.132***	0.536
3	2.349**	−0.544***	0.503	3.370***	−0.572***	−0.059***	−0.171***	0.515	4.597***	−0.588***	−0.056***	−0.212***	−0.115***	0.525
4	1.421	−0.537***	0.503	2.534**	−0.567***	−0.028***	−0.198***	0.517	3.315***	−0.577***	−0.026***	−0.224***	−0.073***	0.521
5	2.445**	−0.533***	0.507	3.516***	−0.561***	−0.004	−0.199***	0.522	4.303***	−0.572***	−0.002	−0.225***	−0.073***	0.526
6	1.560	−0.508***	0.510	2.029*	−0.519***	0.048***	−0.104***	0.516	2.954***	−0.531***	0.051***	−0.135***	−0.086***	0.522
7	1.823*	−0.521***	0.515	2.044*	−0.525***	0.045***	−0.056***	0.518	2.907***	−0.537***	0.047***	−0.085***	−0.081***	0.523
8	2.685**	−0.569***	0.539	2.643**	−0.566***	0.058***	−0.012	0.541	3.726***	−0.580***	0.061***	−0.048***	−0.101***	0.548
9	4.072***	−0.621***	0.537	3.788***	−0.612***	0.055***	0.034***	0.538	4.690***	−0.624***	0.057***	0.004	−0.084***	0.543
High	6.482***	−0.764***	0.543	5.179***	−0.732***	−0.066***	0.266***	0.559	5.436***	−0.735***	−0.066***	0.257***	−0.024***	0.559

Panel F: Volatility-sorted portfolios														
High	14.424***	−0.552***	0.353	16.184***	−0.611***	−0.532***	−0.148***	0.437	16.842***	−0.620***	−0.530***	−0.170***	−0.061***	0.439
2	15.543***	−0.649***	0.462	17.245***	−0.706***	−0.502***	−0.147***	0.533	17.762***	−0.713***	−0.500***	−0.164***	−0.048***	0.534
3	13.926***	−0.630***	0.480	15.648***	−0.686***	−0.451***	−0.168***	0.546	16.359***	−0.695***	−0.449***	−0.192***	−0.066***	0.548
4	10.830***	−0.584***	0.485	12.536***	−0.638***	−0.379***	−0.190***	0.543	13.134***	−0.645***	−0.377***	−0.210***	−0.056***	0.545
5	9.686***	−0.535***	0.485	11.328***	−0.586***	−0.306***	−0.202***	0.535	11.932***	−0.593***	−0.305***	−0.222***	−0.056***	0.537
6	7.879***	−0.498***	0.490	9.365***	−0.543***	−0.238***	−0.196***	0.530	9.941***	−0.551***	−0.237***	−0.216***	−0.054***	0.533
7	7.269***	−0.442***	0.485	8.376***	−0.475***	−0.164***	−0.151***	0.511	9.055***	−0.484***	−0.162***	−0.174***	−0.063***	0.515
8	7.128***	−0.384***	0.469	8.027***	−0.410***	−0.111***	−0.130***	0.488	8.705***	−0.419***	−0.109***	−0.153***	−0.063***	0.494
9	6.555***	−0.303***	0.414	7.321***	−0.325***	−0.071***	−0.119***	0.432	7.860***	−0.332***	−0.070***	−0.137***	−0.050***	0.437
Low	5.501***	−0.173***	0.244	5.777***	−0.180***	−0.021***	−0.044***	0.248	6.290***	−0.187***	−0.019***	−0.062***	−0.048***	0.256
Panel G: Industry-sorted portfolios														
NoDur	3.639***	−0.433***	0.435	3.521***	−0.428***	0.066***	−0.001	0.438	4.677***	−0.443***	0.069***	−0.039***	−0.108***	0.450
Durbl	4.530***	−0.675***	0.464	5.659***	−0.706***	−0.037***	−0.198***	0.473	6.180***	−0.713***	−0.036***	−0.216***	−0.049***	0.474
Manuf	5.373***	−0.592***	0.543	5.768***	−0.602***	−0.019**	−0.067***	0.545	6.824***	−0.616***	−0.017*	−0.102***	−0.099***	0.551
Enrgy	0.412	−0.554***	0.338	1.041	−0.568***	0.119***	−0.158***	0.349	2.543	−0.587***	0.123***	−0.208***	−0.140***	0.358
HiTec	6.708***	−0.784***	0.490	4.253***	−0.721***	−0.043***	0.473***	0.529	4.154***	−0.719***	−0.043***	0.476***	0.009	0.529
Telcm	1.287	−0.539***	0.413	1.678	−0.546***	0.113***	−0.111***	0.422	1.773	−0.548***	0.113***	−0.115***	−0.009	0.422
Shops	5.005***	−0.550***	0.467	4.834***	−0.546***	−0.030***	0.042***	0.468	5.621***	−0.556***	−0.028***	0.016	−0.074***	0.472
Hlth	3.262**	−0.524***	0.414	2.382*	−0.499***	0.093***	0.133***	0.422	3.497***	−0.513***	0.096***	0.095***	−0.104***	0.429
Utils	4.028***	−0.348***	0.299	5.111***	−0.375***	0.067***	−0.225***	0.331	5.964***	−0.386***	0.069***	−0.254***	−0.080***	0.338
Other	6.459***	−0.649***	0.536	8.021***	−0.693***	−0.135***	−0.246***	0.557	8.185***	−0.695***	−0.134***	−0.251***	−0.015*	0.557

Table 4.15 Trading intensity with zero cash rate
This table reports the results for the improvement delivered by the MA switching strategy over the buy-and-hold strategy with a zero cash rate, the trading frequency as well as the BETC using ten decile portfolios sorted by various characteristics. The sample period covers January 4, 1960 until December 31, 2013 with value-weighted portfolio returns. $\Delta\mu$ is the annualized improvement in the average in-sample monthly return, $\Delta\sigma$ is the annualized improvement in the return standard deviation, p_A is the proportion of months during which there is a hold signal, NT is the number of transactions (buy or sell) over the entire sample period, BETC is the break-even one-sided transaction cost in percent, p_1 is the proportion of months during which a buy signal was followed by a positive return of the underlying portfolio, and p_2 is the proportion of months during which a buy signal was followed by a portfolio return in excess of the risk-free rate. The length of the MA window is 20 days.

Portfolio	$\Delta\mu$	$\Delta\sigma$	p_A	NT	BETC	p_1	p_2
			Panel A: Size-sorted portfolios				
Low	11.91	5.22	0.60	931	0.69	39.37	57.78
2	8.19	6.18	0.61	1071	0.41	37.57	56.44
3	6.72	6.12	0.61	1089	0.34	37.34	56.48
4	6.05	6.00	0.62	1147	0.29	37.17	56.08
5	4.54	5.91	0.62	1139	0.22	37.07	56.05
6	4.40	5.59	0.62	1165	0.21	37.13	55.84
7	4.08	5.65	0.62	1145	0.19	36.86	55.50
8	2.65	5.70	0.62	1193	0.12	36.06	54.99
9	0.22	5.71	0.62	1313	0.01	34.89	54.43
High	−3.16	5.85	0.61	1501	−0.11	32.20	52.85
			Panel B: Book-to-market sorted portfolios				
Low	0.88	6.59	0.59	1409	0.03	32.08	52.84
2	−0.04	5.82	0.61	1339	−0.00	33.30	53.19
3	0.29	5.49	0.61	1345	0.01	33.38	53.07
4	0.66	5.58	0.61	1315	0.03	33.98	53.69
5	−1.63	5.51	0.61	1407	−0.06	33.15	53.15
6	−0.36	5.32	0.62	1259	−0.02	34.69	53.77
7	−0.47	5.14	0.62	1301	−0.02	34.74	53.95
8	−0.71	5.55	0.63	1315	−0.03	34.88	54.23
9	−0.21	5.50	0.63	1309	−0.01	35.11	54.11
High	1.75	6.04	0.62	1281	0.07	34.80	53.97

continued

Table 4.15 Continued

Portfolio	$\Delta\mu$	$\Delta\sigma$	p_A	NT	BETC	p_1	p_2
			Panel C: Momentum-sorted portfolios				
Low	13.78	9.07	0.52	1197	0.62	28.87	50.78
2	6.39	7.08	0.56	1295	0.27	30.27	51.25
3	3.50	5.90	0.58	1315	0.14	31.57	52.40
4	1.60	5.75	0.59	1359	0.06	31.83	52.55
5	−0.17	5.41	0.60	1361	−0.01	32.74	52.98
6	−0.61	5.32	0.62	1401	−0.02	33.93	53.48
7	−0.93	5.40	0.62	1355	−0.04	34.10	53.58
8	−2.13	5.37	0.63	1389	−0.08	35.25	54.65
9	−1.26	6.05	0.62	1435	−0.05	34.69	54.20
High	0.16	7.72	0.64	1307	0.01	36.90	55.74
			Panel D: Short-term reversal sorted portfolios				
Low	−7.21	10.43	0.72	1241	−0.32	43.12	57.69
2	−1.69	8.67	0.64	1277	−0.07	36.68	54.68
3	−0.82	7.39	0.62	1345	−0.03	34.83	53.64
4	0.95	6.56	0.61	1299	0.04	33.96	53.37
5	0.31	6.09	0.61	1355	0.01	33.86	53.45
6	0.11	5.48	0.62	1315	0.00	34.24	53.56
7	−0.06	5.22	0.60	1413	−0.00	32.65	53.13
8	1.04	4.86	0.58	1401	0.04	31.45	52.97
9	0.31	4.82	0.57	1409	0.01	30.86	53.08
High	8.17	6.28	0.47	1475	0.30	24.24	50.48
			Panel E: Long-term reversal sorted portfolios				
Low	4.55	6.96	0.58	1295	0.19	32.68	53.33
2	0.33	5.96	0.61	1309	0.01	33.90	53.92
3	−0.93	5.32	0.62	1429	−0.04	34.08	53.53
4	−1.82	5.30	0.62	1389	−0.07	33.87	53.57
5	−0.77	5.24	0.62	1363	−0.03	34.01	53.53
6	−1.51	4.98	0.63	1311	−0.06	34.11	53.51
7	−1.32	5.06	0.62	1407	−0.05	34.08	53.95
8	−0.75	5.65	0.61	1397	−0.03	33.12	53.05
9	0.32	6.29	0.60	1425	0.01	32.20	52.84
High	1.87	7.92	0.59	1375	0.07	32.85	53.36

continued

Table 4.15 Continued

Portfolio	$\Delta\mu$	$\Delta\sigma$	p_A	NT	BETC	p_1	p_2
			Panel F: Volatility-sorted portfolios				
High	11.09	6.18	0.65	935	0.64	41.24	57.91
2	11.62	7.04	0.61	1097	0.58	37.18	56.13
3	10.12	6.74	0.61	1041	0.53	37.69	56.20
4	7.31	6.17	0.62	1053	0.38	38.64	56.70
5	6.46	5.64	0.63	1025	0.34	39.20	57.10
6	4.87	5.22	0.64	1047	0.25	39.39	57.05
7	4.60	4.69	0.65	1091	0.23	40.15	57.61
8	4.81	4.22	0.66	989	0.26	40.99	57.72
9	4.72	3.64	0.67	927	0.28	42.44	58.60
Low	4.46	2.73	0.68	831	0.29	44.71	59.99
			Panel G: Industry-sorted portfolios				
NoDur	1.02	4.58	0.63	1299	0.04	35.53	54.62
Durbl	0.45	6.88	0.57	1391	0.02	30.01	51.15
Manuf	1.80	5.81	0.61	1315	0.07	33.52	53.22
Enrgy	−2.93	6.54	0.59	1431	−0.11	31.72	52.46
HiTec	1.97	8.27	0.57	1407	0.08	30.80	52.39
Telcm	−1.97	5.87	0.58	1461	−0.07	30.10	51.28
Shops	1.68	5.69	0.60	1313	0.07	33.74	53.47
Hlth	0.10	5.72	0.60	1331	0.00	32.76	53.13
Utils	1.93	4.59	0.61	1213	0.09	34.75	53.87
Other	2.54	6.54	0.61	1273	0.11	34.19	53.94

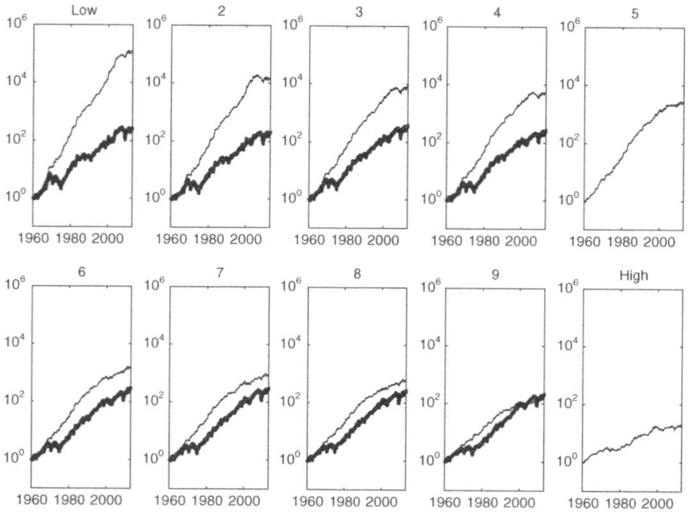

Figure 4.22 Cumulative returns of BH versus MA strategy with skipping a day: Size decile portfolios.

Table 4.16 Market timing with zero cash rate

This table reports alphas, betas, and adjusted R^2 of the market-timing regressions of the MAP excess returns on the market factor using portfolios sorted by various characteristics using a zero cash rate. The TM panel reports the results using the Treynor and Mazuy (1966) quadratic regression with the squared market factor (β_{m^2}) while the HM panel reports the results using the Merton and Henriksson (1981) regression with option-like returns on the market (γ_m). The sample period covers January 4, 1960 until December 31, 2013 with value-weighted portfolio returns. The length of the MA window is 20 days. Newey and West (1987) standard errors with five lags are used in reporting statistical significance of a two-sided null hypothesis at the 1%, 5%, and 10% level given by a***, a**, and a*, respectively.

Portfolio	α	β_m	β_{m^2}	$\overline{R}^2$	α	β_m	γ_m	$\overline{R}^2$
			Panel A: Size-sorted portfolios					
Low	11.319***	−0.429***	0.013***	0.412	2.208	−0.357***	0.149***	0.412
2	10.390***	−0.559***	0.005***	0.454	5.110***	−0.520***	0.078***	0.455
3	8.288***	−0.571***	0.008***	0.484	2.296	−0.525***	0.096***	0.484
4	8.486***	−0.570***	0.004***	0.495	5.194***	−0.545***	0.052***	0.495
5	7.606***	−0.575***	0.002	0.511	5.829***	−0.562***	0.027*	0.511
6	6.780***	−0.551***	0.004***	0.529	4.395***	−0.532***	0.041***	0.529
7	6.442***	−0.563***	0.004***	0.540	4.500***	−0.546***	0.036***	0.540
8	4.977***	−0.580***	0.005***	0.560	2.678*	−0.561***	0.042***	0.560
9	1.427	−0.586***	0.010***	0.585	0.174	−0.567***	0.044***	0.583
High	−1.382	−0.608***	0.008***	0.590	−1.284	−0.600***	0.022*	0.588

continued

Table 4.17 Continued

Portfolio	α	β_m	β_{m^2}	$\overline{R}^2$	α	β_m	γ_m	$\overline{R}^2$
Panel B: Book-to-market sorted portfolios								
Low	4.231***	−0.661***	0.003**	0.564	4.698***	−0.661***	0.002	0.563
2	2.321**	−0.597***	0.005***	0.565	2.377	−0.592***	0.015	0.564
3	1.623	−0.558***	0.008***	0.548	0.141	−0.539***	0.043***	0.546
4	1.133	−0.568***	0.012***	0.550	−1.012	−0.541***	0.062***	0.547
5	0.620	−0.551***	0.004***	0.520	1.897	−0.554***	−0.002	0.519
6	1.311	−0.531***	0.006***	0.519	1.020	−0.523***	0.022*	0.517
7	0.503	−0.509***	0.009***	0.502	−0.252	−0.495***	0.035***	0.500
8	0.191	−0.526***	0.009***	0.471	−0.430	−0.512***	0.035***	0.469
9	0.807	−0.536***	0.009***	0.484	−0.022	−0.521***	0.037***	0.482
High	3.104**	−0.577***	0.009***	0.459	1.501	−0.558***	0.046***	0.458
Panel C: Momentum-sorted portfolios								
Low	20.294***	−0.833***	−0.006***	0.451	21.539***	−0.848***	−0.033	0.451
2	11.646***	−0.699***	−0.004***	0.494	12.186***	−0.707***	−0.019	0.493
3	7.441***	−0.605***	−0.001	0.510	6.754***	−0.602***	0.005	0.510
4	3.526***	−0.588***	0.007***	0.534	1.772	−0.570***	0.041***	0.533
5	3.242***	−0.559***	−0.000	0.527	4.921***	−0.570***	−0.021*	0.527
6	1.368	−0.538***	0.005***	0.521	3.273**	−0.544***	−0.008	0.521
7	0.488	−0.543***	0.008***	0.533	0.322	−0.533***	0.025**	0.532
8	−1.737	−0.536***	0.012***	0.516	−3.226**	−0.514***	0.053***	0.513
9	−0.149	−0.591***	0.010***	0.527	−2.633	−0.564***	0.060***	0.525
High	2.039	−0.703***	0.010***	0.475	0.114	−0.681***	0.052***	0.474

Panel D: Short-term reversal sorted portfolios								
Low	0.785	−0.837***	−0.012***	0.392	10.385***	−0.911***	−0.152***	0.393
2	3.669**	−0.757***	−0.003**	0.495	7.577***	−0.785***	−0.057***	0.495
3	3.805***	−0.669***	−0.002*	0.509	7.100***	−0.693***	−0.047***	0.510
4	3.255***	−0.620***	0.006***	0.534	4.633***	−0.623***	0.001	0.533
5	1.982*	−0.593***	0.008***	0.552	1.443	−0.581***	0.030**	0.551
6	1.211	−0.556***	0.009***	0.541	0.435	−0.542***	0.037***	0.539
7	1.470	−0.553***	0.008***	0.559	0.726	−0.541***	0.031***	0.557
8	2.332**	−0.536***	0.008***	0.548	−1.197	−0.505***	0.067***	0.547
9	1.934*	−0.543***	0.007***	0.530	−0.857	−0.518***	0.054***	0.529
High	10.480***	−0.657***	0.007***	0.525	2.117	−0.598***	0.122***	0.527
Panel E: Long-term reversal sorted portfolios								
Low	5.823***	−0.673***	0.012***	0.485	2.407	−0.639***	0.076***	0.484
2	1.436	−0.592***	0.010***	0.512	1.530	−0.581***	0.029**	0.509
3	−0.118	−0.539***	0.010***	0.506	−1.137	−0.522***	0.042***	0.504
4	−1.606	−0.531***	0.012***	0.507	−1.736	−0.517***	0.038***	0.503
5	−0.892	−0.527***	0.014***	0.512	−1.637	−0.507***	0.049***	0.507
6	0.321	−0.506***	0.005***	0.510	0.648	−0.502***	0.011	0.510
7	0.318	−0.518***	0.006***	0.517	−1.040	−0.503***	0.035***	0.516
8	0.652	−0.565***	0.008***	0.541	−1.581	−0.542***	0.052***	0.540
9	3.666***	−0.620***	0.002	0.537	2.624	−0.612***	0.018	0.537
High	7.468***	−0.766***	−0.004***	0.543	7.904***	−0.773***	−0.017	0.543

continued

Table 4.17 Continued

Portfolio	α	β_m	β_{m^2}	$\bar{R}^2$	α	β_m	γ_m	$\bar{R}^2$
Panel F: Volatility-sorted portfolios								
High	11.847***	−0.547***	0.011***	0.355	4.728**	−0.491***	0.117***	0.355
2	12.756***	−0.644***	0.011***	0.465	4.622**	−0.580***	0.132***	0.465
3	12.573***	−0.627***	0.006***	0.481	8.306***	−0.594***	0.068***	0.481
4	9.635***	−0.581***	0.005***	0.486	6.800***	−0.558***	0.049***	0.486
5	8.482***	−0.533***	0.005***	0.486	4.838***	−0.505***	0.059***	0.486
6	6.509***	−0.496***	0.006***	0.491	4.052**	−0.474***	0.046***	0.490
7	5.675***	−0.439***	0.007***	0.486	2.951**	−0.415***	0.052***	0.485
8	4.481***	−0.379***	0.011***	0.475	0.338	−0.341***	0.082***	0.472
9	3.951***	−0.298***	0.011***	0.423	−0.848	−0.257***	0.090***	0.419
Low	3.176***	−0.168***	0.010***	0.257	−0.616	−0.134***	0.074***	0.250
Panel G: Industry-sorted portfolios								
NoDur	1.092	−0.428***	0.010***	0.440	−1.970	−0.398***	0.068***	0.437
Durbl	2.885*	−0.672***	0.007***	0.465	3.194	−0.667***	0.016	0.464
Manuf	2.984**	−0.587***	0.010***	0.546	1.659	−0.568***	0.045***	0.543
Enrgy	−0.562	−0.552***	0.004***	0.338	0.977	−0.558***	−0.007	0.338
HiTec	6.074***	−0.783***	0.003*	0.490	7.663***	−0.790***	−0.012	0.490
Telcm	0.959	−0.538***	0.001	0.413	3.530**	−0.553***	−0.027*	0.413
Shops	3.756***	−0.548***	0.005***	0.468	3.851**	−0.543***	0.014	0.467
Hlth	1.113	−0.520***	0.009***	0.416	0.527	−0.507***	0.033**	0.414
Utils	3.755***	−0.348***	0.001	0.300	3.182**	−0.343***	0.010	0.300
Other	5.199***	−0.646***	0.005***	0.537	3.000*	−0.627***	0.042***	0.537

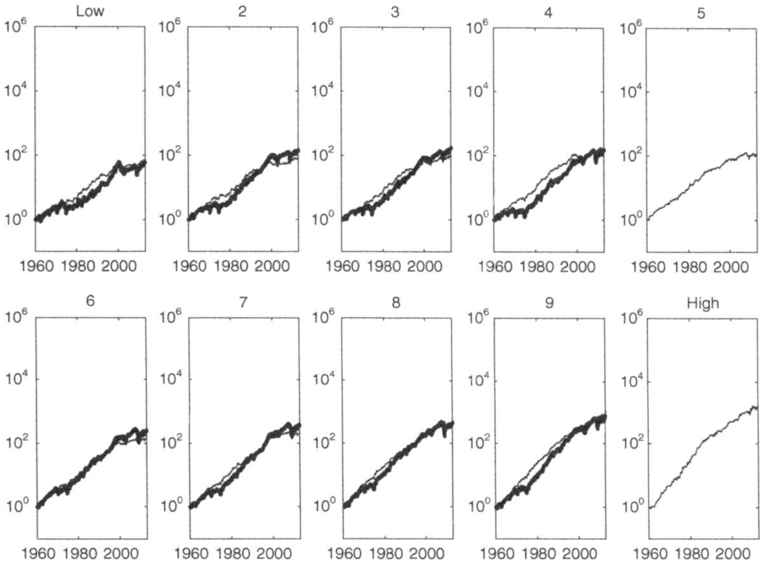

Figure 4.23 Cumulative returns of BH versus MA strategy with skipping a day: Book-to-market decile portfolios.

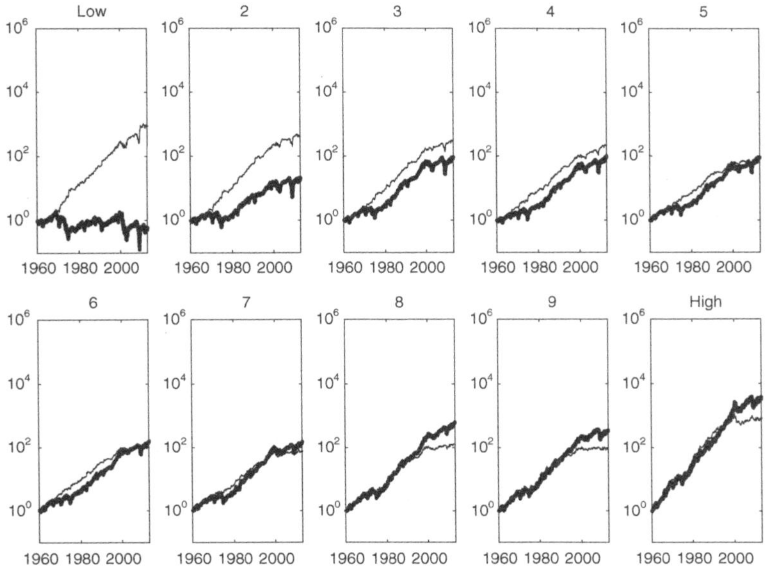

Figure 4.24 Cumulative returns of BH versus MA strategy with skipping a day: Momentum decile portfolios.

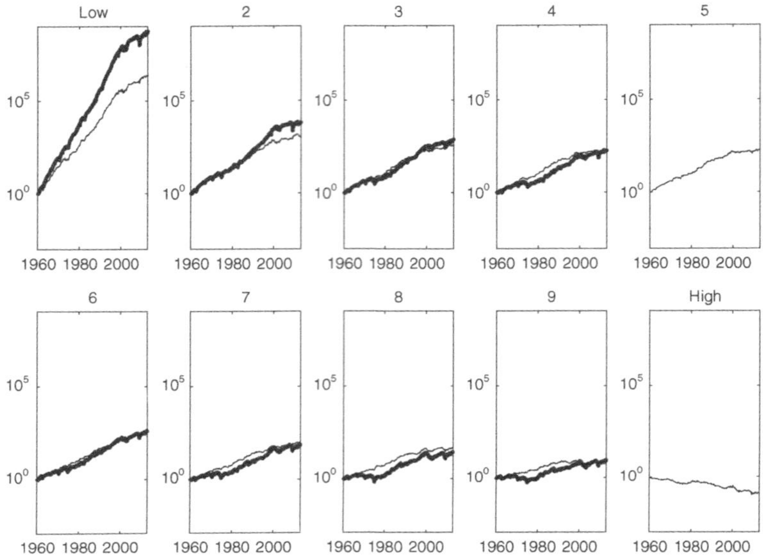

Figure 4.25 Cumulative returns of BH versus MA strategy with skipping a day: Short-term reversal decile portfolios.

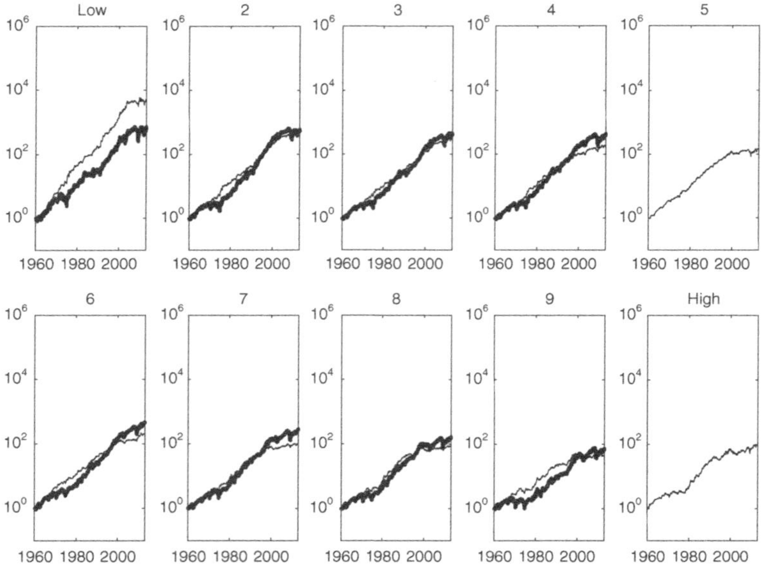

Figure 4.26 Cumulative returns of BH versus MA strategy with skipping a day: Long-term reversal decile portfolios.

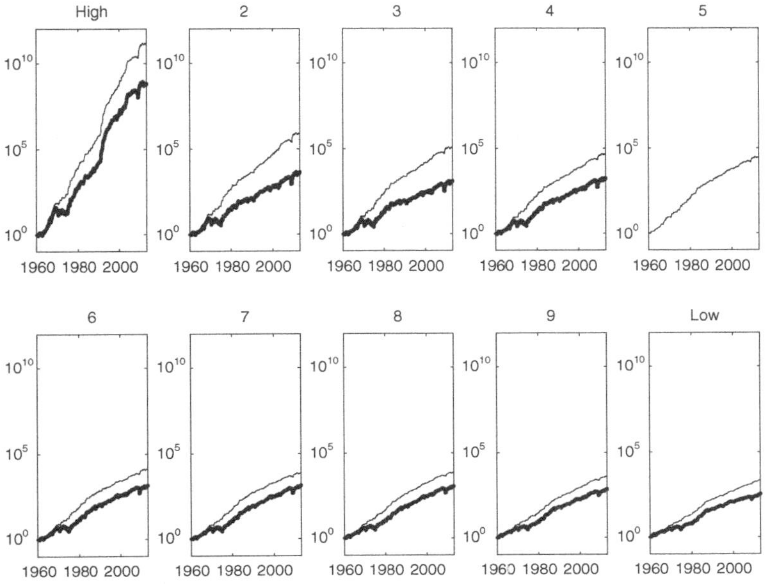

Figure 4.27 Cumulative returns of BH versus MA strategy with skipping a day: Volatility decile portfolios.

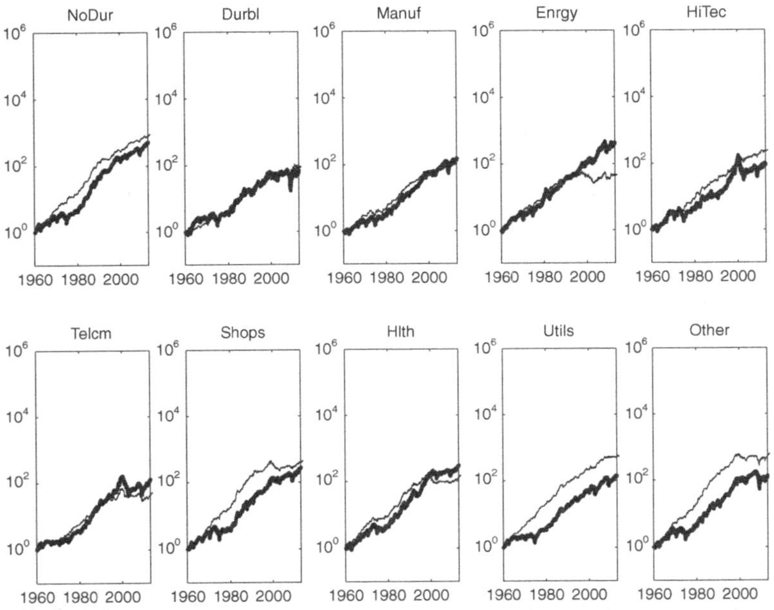

Figure 4.28 Cumulative returns of BH versus MA strategy with skipping a day: Industry portfolios.

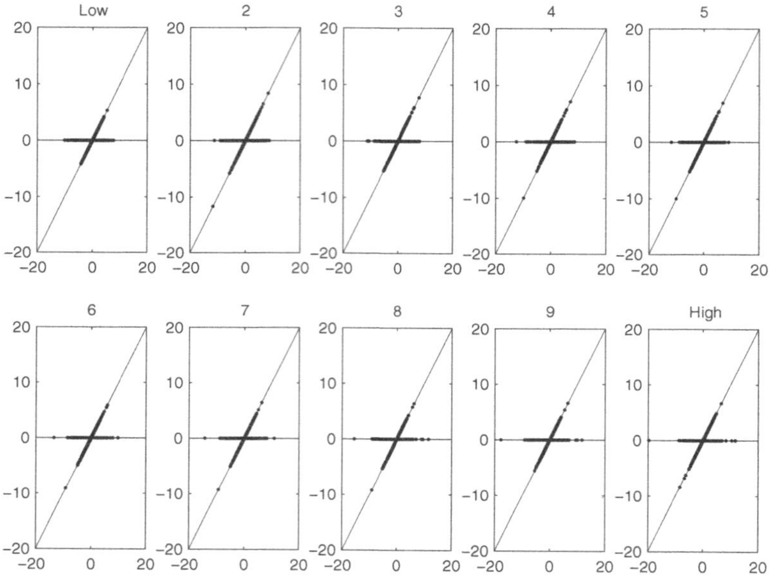

Figure 4.29 Scatter plot of BH versus MA returns with a zero cash rate: Size decile portfolios.

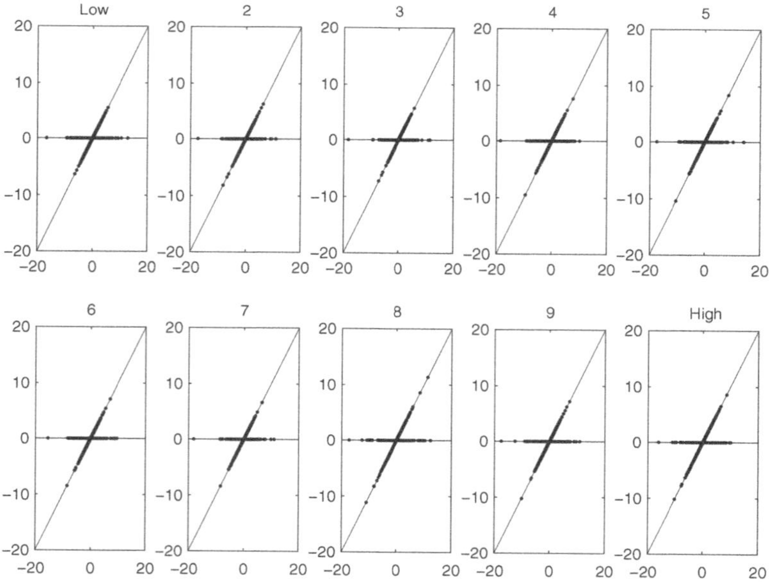

Figure 4.30 Scatter plot of BH versus MA returns with a zero cash rate: Book-to-market decile portfolios.

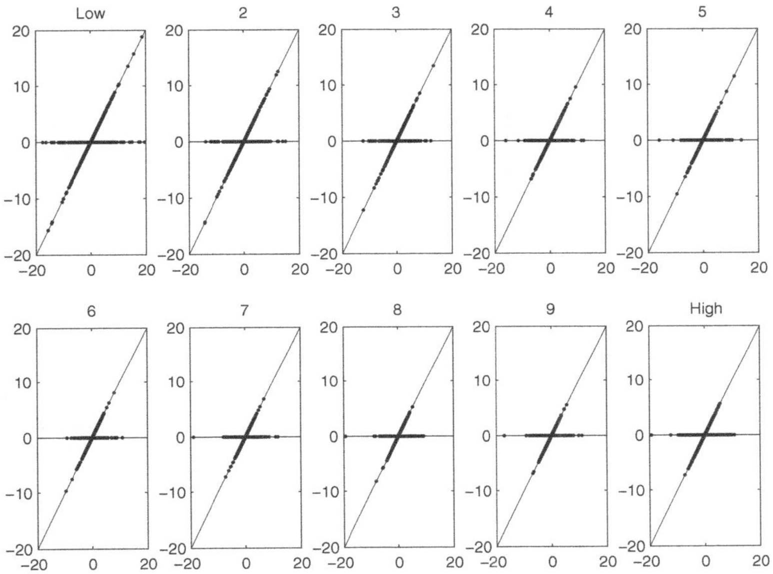

Figure 4.31 Scatter plot of BH versus MA returns with a zero cash rate: Momentum decile portfolios.

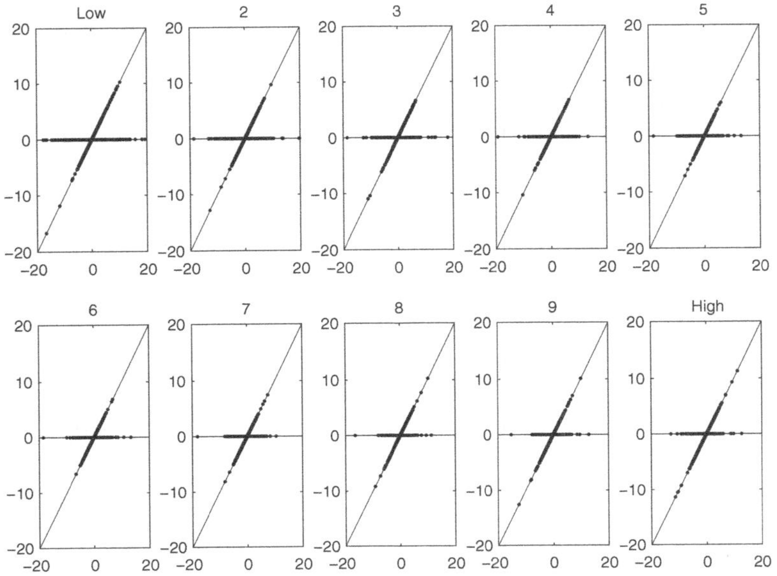

Figure 4.32 Scatter plot of BH versus MA returns with a zero cash rate: Short-term reversal decile portfolios.

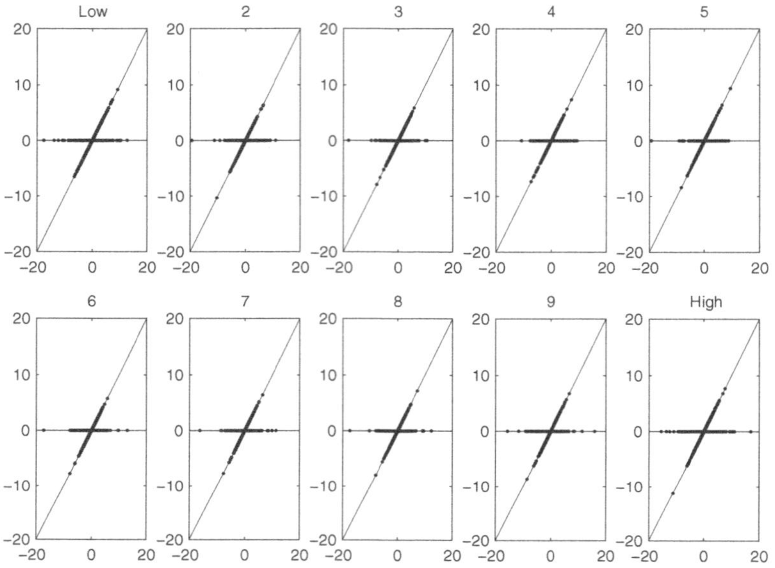

Figure 4.33 Scatter plot of BH versus MA returns with a zero cash rate: Long-term reversal decile portfolios.

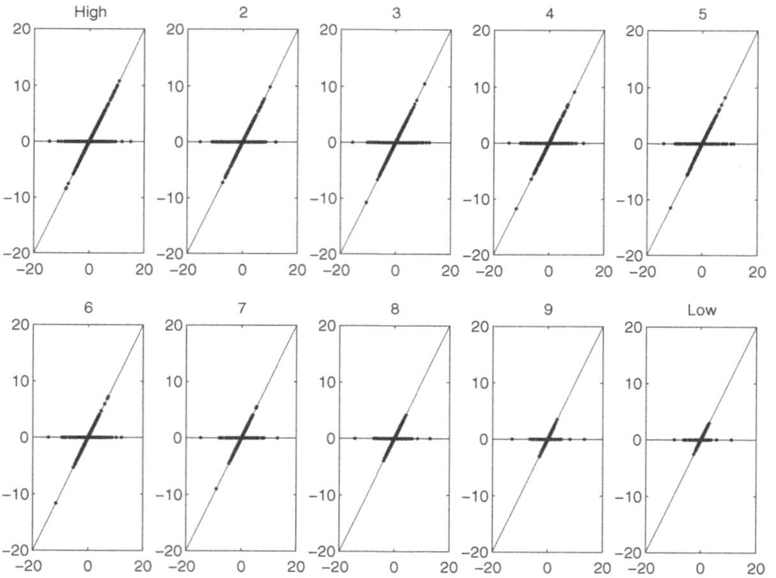

Figure 4.34 Scatter plot of BH versus MA returns with a zero cash rate: Volatility decile portfolios.

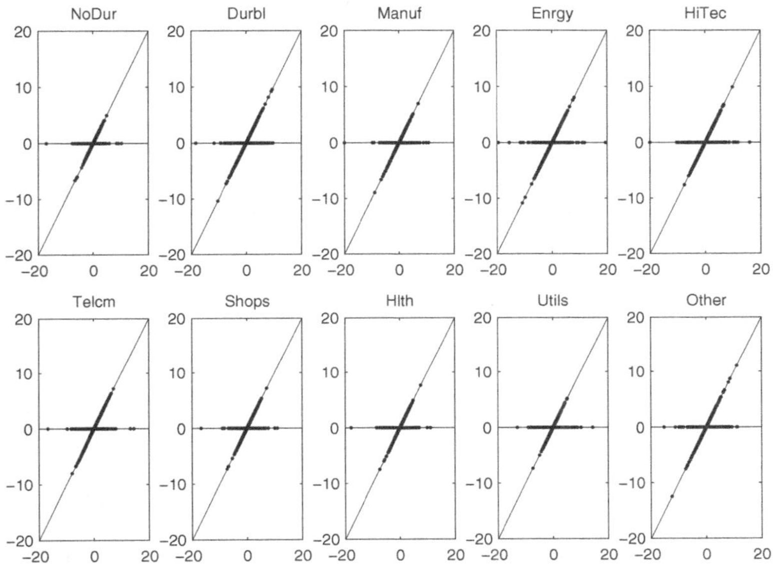

Figure 4.35 Scatter plot of BH versus MA returns with a zero cash rate: Industry portfolios.

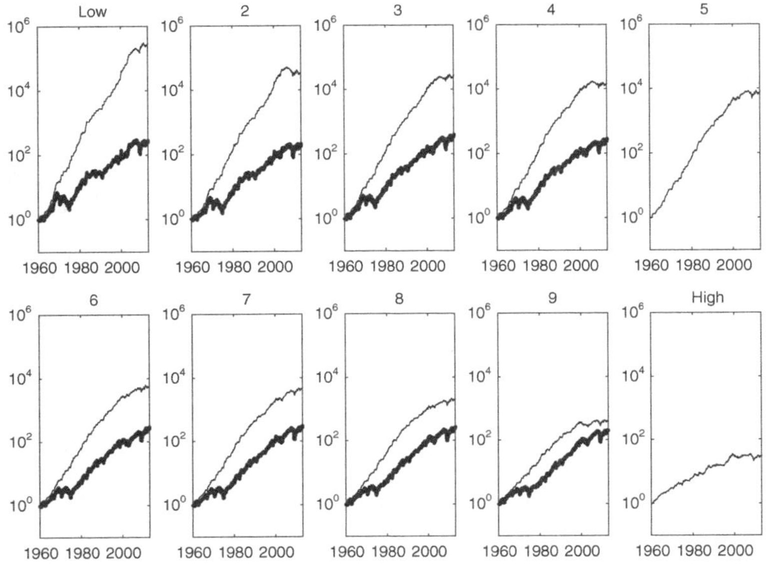

Figure 4.36 Cumulative returns of BH versus MA strategy with a zero cash rate: Size decile portfolios.

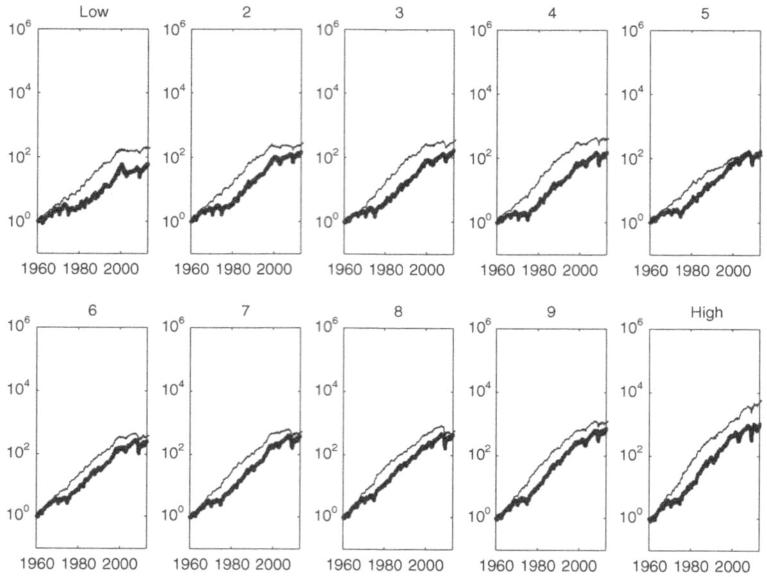

Figure 4.37 Cumulative returns of BH versus MA strategy with a zero cash rate: Book-to-market decile portfolios.

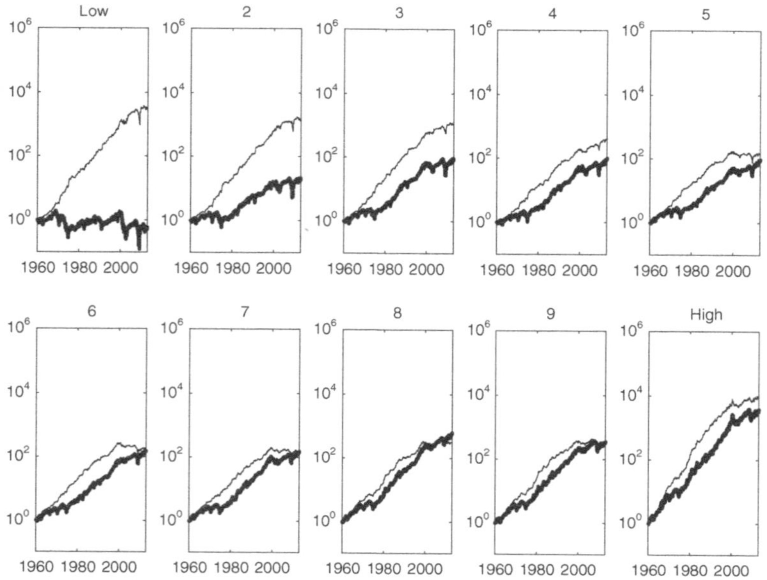

Figure 4.38 Cumulative returns of BH versus MA strategy with a zero cash rate: Momentum decile portfolios.

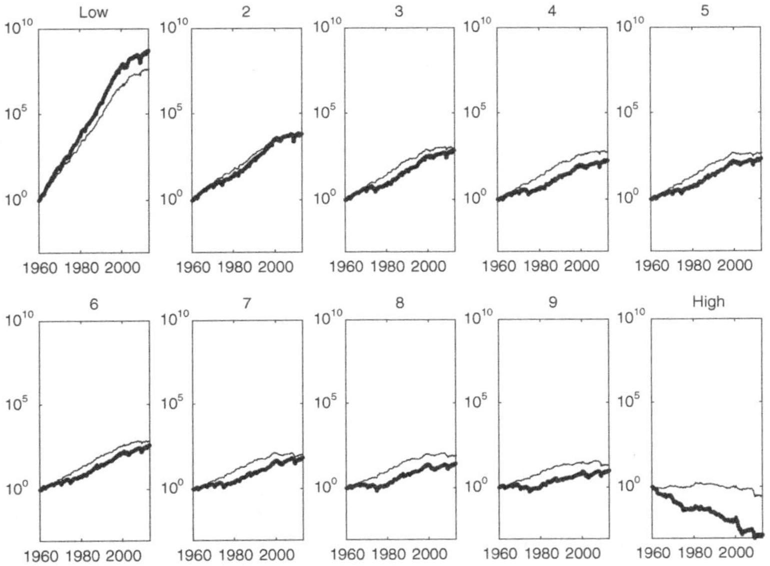

Figure 4.39 Cumulative returns of BH versus MA strategy with a zero cash rate: Short-term reversal decile portfolios.

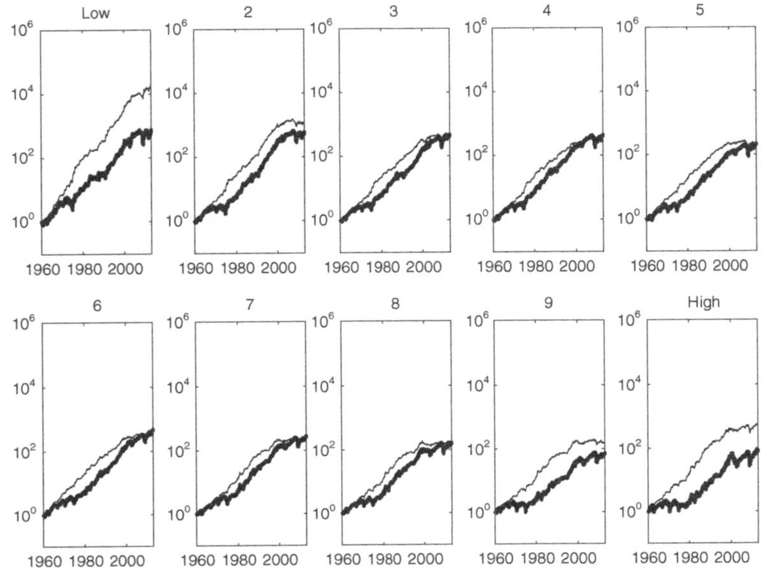

Figure 4.40 Cumulative returns of BH versus MA strategy with a zero cash rate: Long-term reversal decile portfolios.

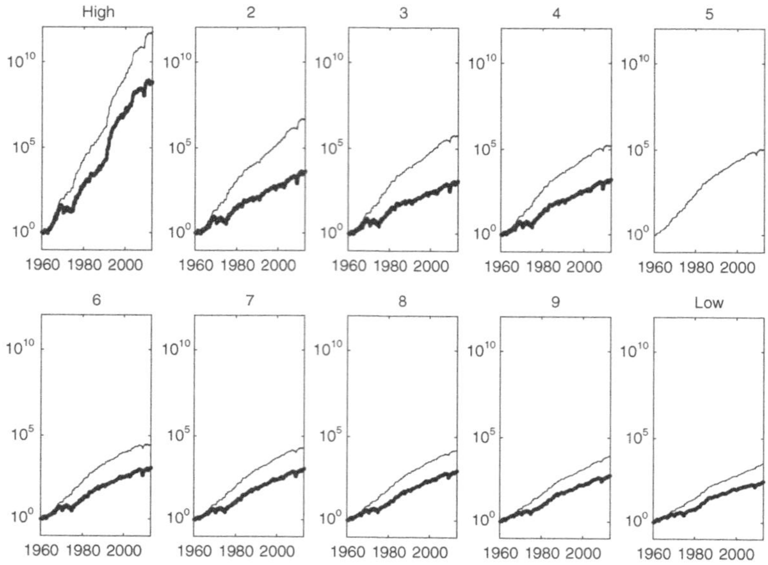

Figure 4.41 Cumulative returns of BH versus MA strategy with a zero cash rate: Volatility decile portfolios.

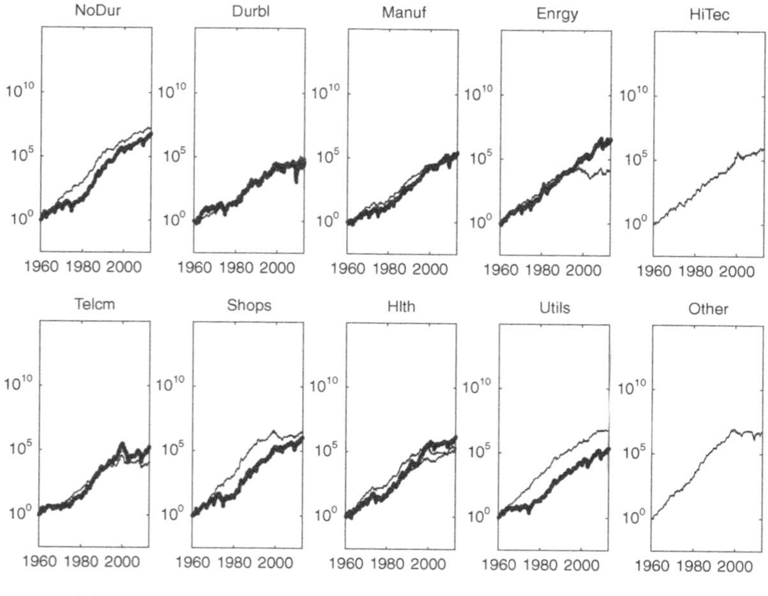

Figure 4.42 Cumulative returns of BH versus MA strategy with a zero cash rate: Industry portfolios.

performance improvement is for the loser momentum portfolio which is 13.8% per year which is a reduction compared to the baseline implementation. Furthermore, when the strategy does not work so well, foregoing the risk-free rate leads to a few portfolios having a negative improvement though the magnitude of the relative loss never exceeds 7% per year.

Finally, Table 4.16 presents the regression estimates for the two most commonly used market timing regressions. The coefficients of primary interest here are β_{m^2} and γ_m. In the vast majority of the cases the estimates for both parameters are positive and statistically significant, indicating that the MA strategy has substantial market timing power. However, a few of the estimates for γ_m are statistically significant leading to the conclusion that there is no evidence of market timing power of the MA strategy for those portfolios.

Figures 4.29 through 4.35 present the scatter plots of the daily returns of the MA strategy versus the daily returns of the BH strategy when we use cash instead of the risk-free asset for the size-sorted deciles, book-to-market sorted deciles, momentum-sorted deciles, short-term reversal deciles, long-term reversal deciles, volatility-sorted deciles, and industry portfolios, respectively.

Figures 4.36 through 4.42 present the cumulative investment performance of $1 invested on January 4, 1960, and continuously reinvested until December 31, 2013, the MA strategy (thin line) and the BH strategy (thick line) when we use cash instead of the risk-free asset for the size-sorted deciles, book-to-market sorted deciles, momentum-sorted deciles, short-term reversal deciles, long-term reversal deciles, volatility-sorted deciles, and industry portfolios, respectively.

Figures 4.22 through 4.28 present the cumulative investment performance of $1 invested on January 4, 1960, and continuously reinvested until December 31, 2013, the MA strategy (thin line) and the BH strategy (thick line) when we skip one day before we trade for the size-sorted deciles, book-to-market sorted deciles, momentum-sorted deciles, short-term reversal deciles, long-term reversal deciles, volatility-sorted deciles, and industry portfolios, respectively.

Chapter 5

Individual Securities

In this chapter, I investigate the performance of the MA strategy relative to the performance of the BH strategy using individual US stocks with various market capitalizations. Specifically, I pick the components of the S&P 500 index as of December 31, 2013, with sufficiently long available price histories to be representative of large capitalization stocks. Similarly, I choose the components of the S&P 400 index as of December 31, 2013, to be representative of mid-cap stocks. Finally, I use the components of the S&P 600 index as representative of small-cap stocks. The historical price data with dividends reinvested is obtained from the Center for Research in Security Prices at the University of Chicago.

5.1 Large-Cap US Stocks

Figure 5.1 presents histogram plots of the frequency distribution of the mean improvement, $\Delta\mu$, the risk improvement, $\Delta\sigma$, the fraction of times the MA strategy is active, p_A, the number of trades, NT, the BETC, as well as the percentage of times the MA strategy return is positive, p_1. The analysis lists the results of the performance of the MA strategy relative to the BH strategy for 483 stocks that were part of the S&P 500 index as of December 31, 2013.[1] The empirical findings indicate that there is a substantial decrease in risk as evidenced by the right-skewed distribution for $\Delta\sigma$. However, the left-skewed distribution for the return improvement, $\Delta\mu$, points out toward a weakness in the MA strategy. Specifically, the

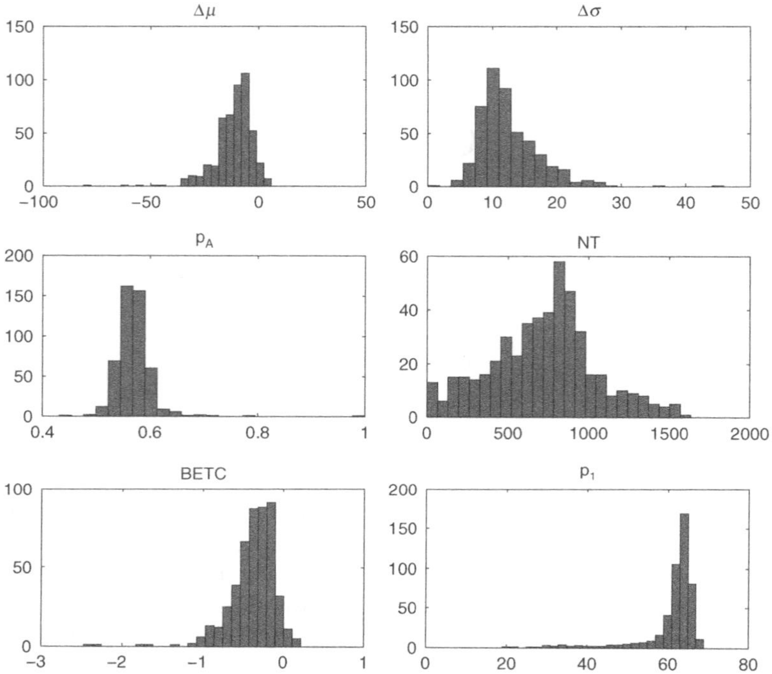

Figure 5.1 Relative performance of BH and MA strategies with large-capitalization US stocks.

large number of S&P 500 components with negative values of $\Delta\mu$ indicate that the MA strategy is inferior to the BH strategy and investors would be better off buying and holding those stocks rather than trying to time the market using an MA. This is also evident when observing the left-skewed distribution for BETC with a large number of stocks with negative values.

Figure 5.2 presents histogram plots of the frequency distribution of the market-timing coefficients of the Treynor–Mazuy market-timing regression, β_{m^2}, and the Henriksson–Merton market timing regression, γ_m, as well as their t-statistics. Both coefficients have negative values for a larger number of stocks than have positive values, indicating that the market-timing ability of the MA strategy fails in this case. This is further confirmed when investigating the statistical significance of the market-timing coefficients by looking at the frequency distribution of their respective t-statistics. A significant number of S&P 500 stocks have negative and statistically significant market-timing coefficients[2] when compared to the number of S&P 500 stocks that have positive and statistically significant

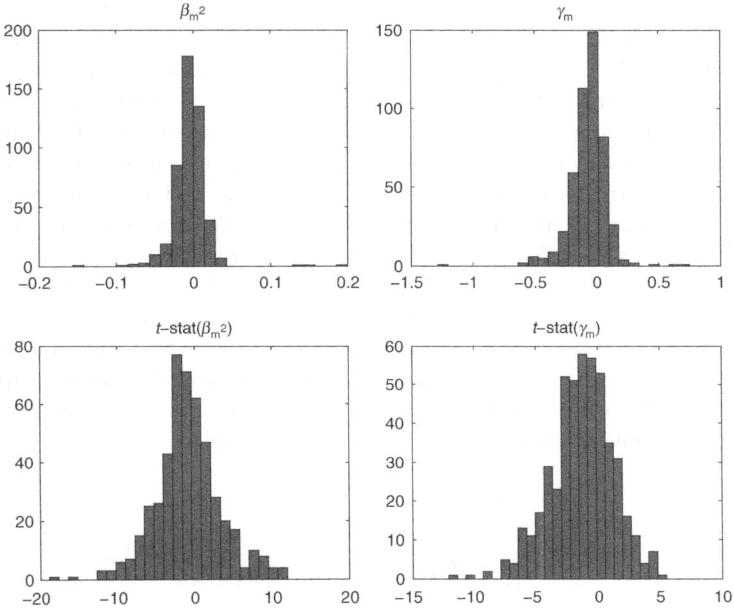

Figure 5.2 Market timing coefficients of MA strategies with large-capitalization US stocks.

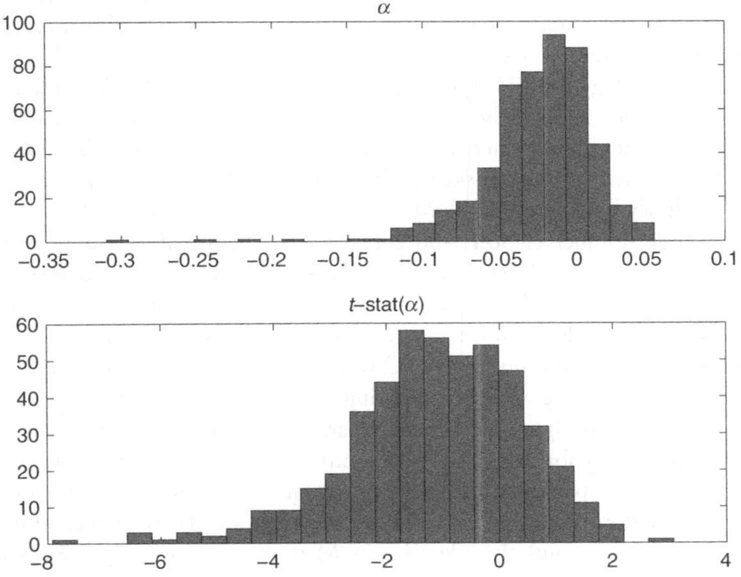

Figure 5.3 Abnormal returns of MA strategies with large-capitalization US stocks.

coefficients.[3] The overall conclusion that can be reached is that more often than not the MA market-timing strategy is on the wrong side of the market.

Finally, Figure 5.3 plots the histograms of the frequency distribution of the abnormal return, α, from the Carhart (1997) four-factor model (top panel) along with their associated t-statistics (bottom panel). Positive and statistically significant returns indicate that a strategy outperforms its benchmark when controlling for factors like market capitalization, book-to-market, and momentum. Vice versa, negative and statistically significant abnormal returns indicate an underperformance. Inspecting the top panel of Figure 5.3 reveals that most of the abnormal returns of the MA strategy for the S&P 500 components are negative. Moreover, a larger number of negative abnormal returns are statistically significant when compared to those that have positive and statistically significant returns. This leads to the conclusion that the MA strategy clearly underperforms the BH strategy when applied to large-cap US stocks.

5.2 Mid-Cap US Stocks

Figure 5.4 presents histogram plots of the frequency distribution of the mean improvement, $\Delta\mu$, the risk improvement, $\Delta\sigma$, the fraction of times the MA strategy is active, p_A, the number of trades, NT, the BETC, as well as the percentage of times the MA strategy return is positive, p_1. The analysis lists the results of the performance of the MA strategy relative to the BH strategy for 374 stocks that were part of the S&P 400 index as of December 31, 2013.[4] The empirical findings indicate once again that there is a significant decrease in risk as evidenced by the right-skewed distribution for $\Delta\sigma$. However, the left-skewed distribution for the return improvement, $\Delta\mu$, clearly indicates that the MA strategy underperforms the BH strategy. The large number of S&P 400 components with negative values of $\Delta\mu$ indicate that the MA strategy is inferior to the BH strategy and investors would be better off buying and holding those stocks rather than trying to time the market using an MA. One can also see that by noting the left-skewed distribution for BETC which has a large number of mid-cap stocks with negative values. Comparing the evidence reported in Figure 5.4 with the one in Figure 5.1 points toward an even weaker performance of the MA strategy for mid-cap stocks compared to that of large-cap stocks.

Figure 5.5 presents histogram plots of the frequency distribution of the market-timing coefficients of the Treynor–Mazuy market-timing regression, β_{m^2}, and the Henriksson–Merton market-timing regression,

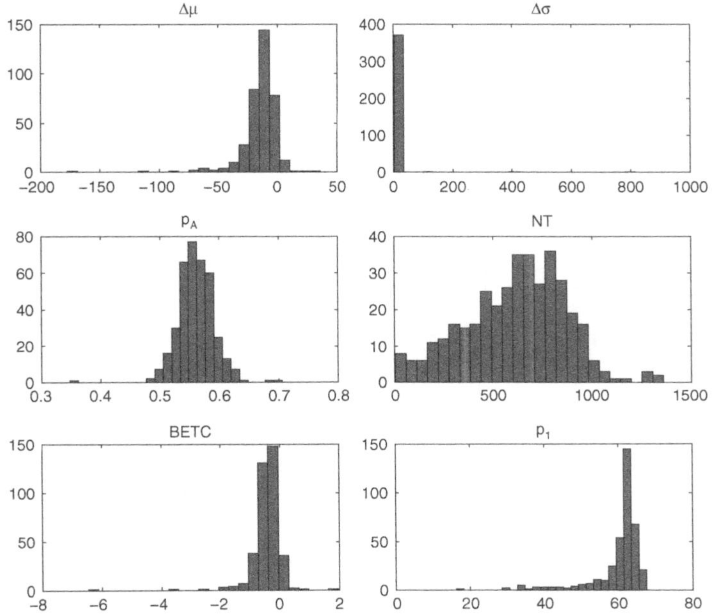

Figure 5.4 Relative performance of BH and MA strategies with mid-capitalization US stocks.

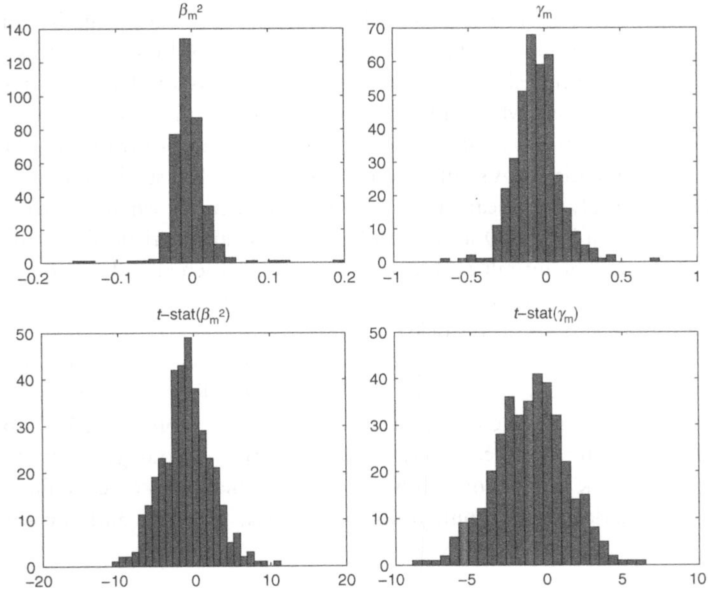

Figure 5.5 Market timing coefficients of MA strategies with mid-capitalization US stocks.

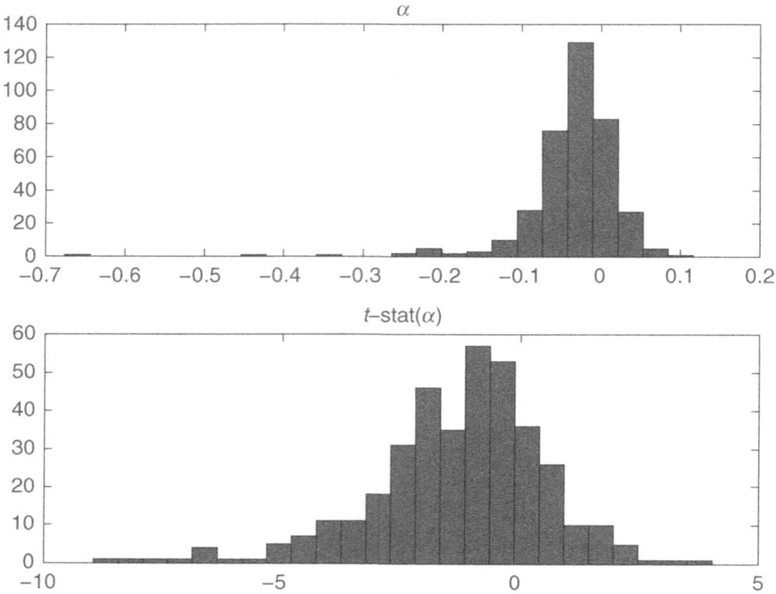

Figure 5.6 Abnormal returns of MA strategies with mid-capitalization US stocks.

γ_m, as well as their *t*-statistics. Both coefficients have negative values for a larger number of stocks than have positive values, indicating that the market-timing ability of the MA strategy fails in this case. This is further confirmed when investigating the statistical significance of the market-timing coefficients by looking at the frequency distribution of their respective *t*-statistics. A significant number of S&P 400 stocks have negative and statistically significant market-timing coefficients when compared to the number of S&P 500 stocks that have positive and statistically significant coefficients. The overall conclusion that can be reached is that more often than not the MA market-timing strategy is on the wrong side of the market when applied to mid-cap stocks.

Figure 5.6 plots the histograms of the frequency distribution of the abnormal return, α, from the Carhart (1997) four-factor model (top panel) along with their associated *t*-statistics (bottom panel). Positive and statistically significant returns indicate that a strategy outperforms its benchmark when controlling for factors like market capitalization, book-to-market, and momentum. Vice versa, negative and statistically significant abnormal returns indicate an underperformance. Inspecting the top panel of Figure 5.6 reveals that most of the abnormal returns of the MA strategy for the S&P 400 components are negative. Moreover, a larger number of negative abnormal returns are statistically significant when compared to those that have positive and statistically significant returns. This leads to the conclusion that the MA strategy clearly underperforms the BH strategy when applied to mid-cap US stocks.

5.3 Small-Cap US Stocks

Figure 5.7 presents histogram plots of the frequency distribution of the mean improvement, $\Delta\mu$, the risk improvement, $\Delta\sigma$, the fraction of times

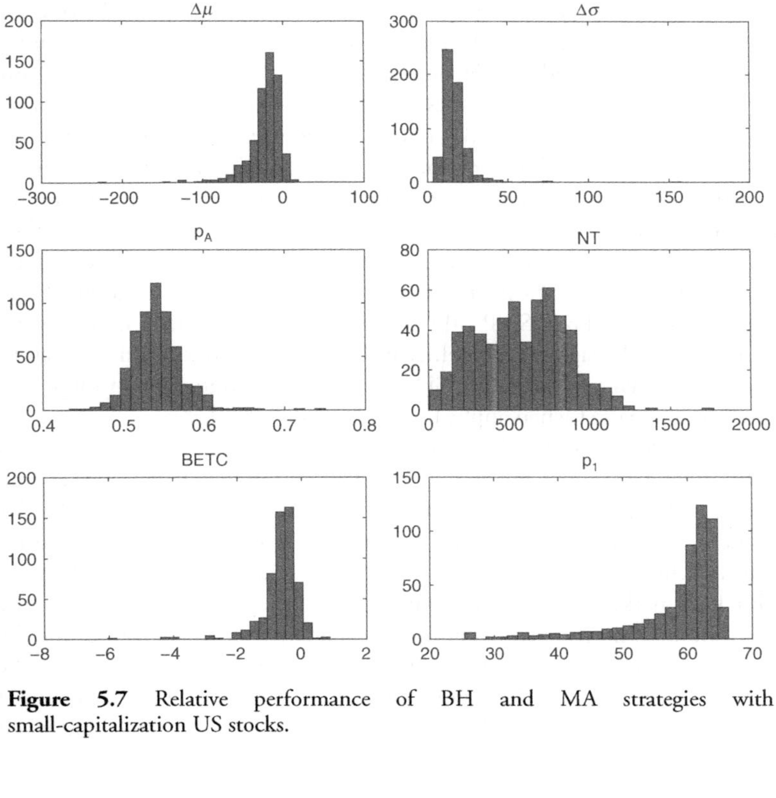

Figure 5.7 Relative performance of BH and MA strategies with small-capitalization US stocks.

the MA strategy is active, p_A, the number of trades, NT, the BETC, as well as the percentage of times the MA strategy return is positive, p_1. The analysis lists the results of the performance of the MA strategy relative to the BH strategy for 571 stocks that were part of the S&P 600 index as of December 31, 2013.[5] The empirical findings indicate once again that there is a significant decrease in risk as evidenced by the right-skewed distribution for $\Delta\sigma$. Once again, the left-skewed distribution for the return improvement, $\Delta\mu$, clearly indicates that the MA strategy underperforms the BH strategy. The large number of S&P 600 components with negative values of $\Delta\mu$ indicate that the MA strategy is inferior to the BH strategy and investors would be better off buying and holding those stocks rather than trying to time the market using an MA. One can also see that by noting the left-skewed distribution for BETC which has a large number of small-cap stocks with negative values. Comparing the evidence reported in Figure 5.1 and 5.4 with that reported in Figure 5.7 points toward an even weaker performance of the MA strategy for small-cap stocks compared to that of both large-cap and mid-cap stocks.

Figure 5.8 presents histogram plots of the frequency distribution of the market-timing coefficients of the Treynor–Mazuy market-timing regression, β_{m^2}, and the Henriksson–Merton market-timing regression, γ_m, as well as their t-statistics. Both coefficients have negative values for a larger number of stocks than have positive values, indicating that the market-timing ability of the MA strategy fails in this case. This is further confirmed when investigating the statistical significance of the market-timing coefficients by looking at the frequency distribution of their respective t-statistics. A significant number of S&P 600 stocks have negative and statistically significant market-timing coefficients when compared to the number of both S&P 500 and S&P 400 stocks that have positive and statistically significant coefficients. The overall conclusion that can be reached is that very often the MA market-timing strategy is on the wrong side of the market when applied to small-cap US stocks.

Figure 5.9 plots the histograms of the frequency distribution of the abnormal return, α, from the Carhart (1997) four-factor model (top panel) along with their associated t-statistics (bottom panel). Positive and statistically significant returns indicate that a strategy outperforms its benchmark when controlling for factors like market capitalization, book-to-market, and momentum. Vice versa, negative and statistically significant abnormal returns indicate an underperformance. Inspecting the top panel of Figure 5.9 reveals that most of the abnormal returns of the MA strategy for the S&P 600 components are negative. Moreover, a larger number of negative abnormal returns are statistically significant when

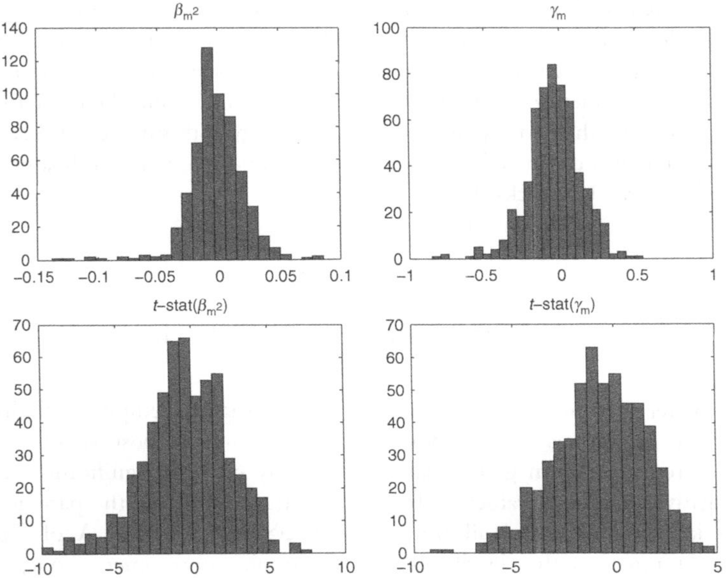

Figure 5.8 Market timing coefficients of MA strategies with small-capitalization US stocks.

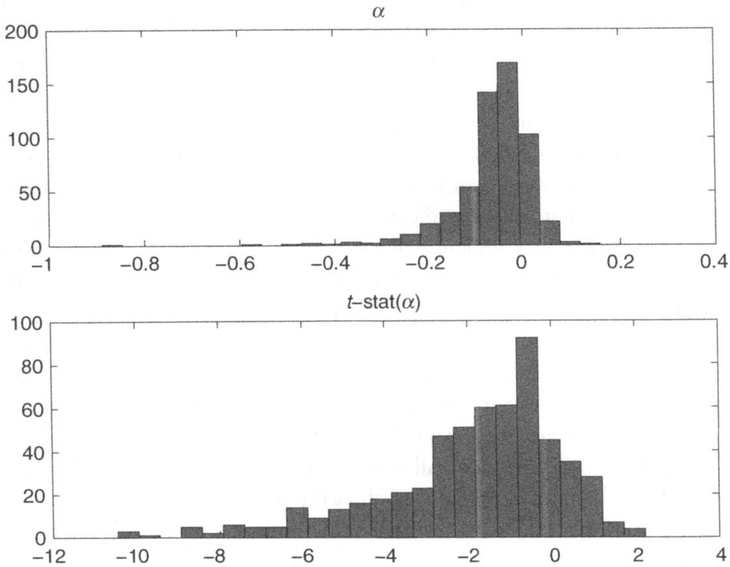

Figure 5.9 Abnormal returns of MA strategies with small-capitalization US stocks.

compared to those that have positive and statistically significant returns. This leads to the conclusion that the MA strategy clearly underperforms the BH strategy when applied to small-cap US stocks. The long left tails of both the frequency distribution for abnormal returns and their t-statistics indicate that the performance of the MA strategy with small-cap US stocks is even weaker compared to the MA strategy performance for both large-cap and mid-cap US stocks.

5.4 Summary for US Stocks

A better and more formal way of summarizing the empirical findings in the histograms is by comparing the frequency of positive values to the frequency of negative values as well as positive significant versus negative significant values. Table 5.1 lists those value for the parameters of key interest across all three sets of US stocks. Panel A of Table 5.1 summarizes the distribution of the mean improvement, $\Delta\mu$, the BETC, the Treynor–Mazuy market-timing regression coefficient, β_{m^2}, the Henriksson–Merton market-timing regression coefficient, γ_m, as well as the four-factor abnormal return, α. First and foremost, it is worthwhile noting the negatively skewed distributions of all of these parameters where both the mean and the median values are negative. This, in and of itself, does not speak much to their statistical significance which is what we turn to next. The next four lines in Panel A of Table 5.1 report the percentage of positive and significant values, $\%PS$, the percentage of positive values regardless of their statistical significance, $\%P$, the percentage of negative values regardless of statistical significance, $\%N$, and the percentage of negative and statistically significant values, $\%NS$. The value for BETC is directly related to $\Delta\mu$ which explains why the cross-sectional distribution and the reported values are identical for both parameters. It is worthwhile noting that negative values outnumber positive values by a factor of almost 30 to 1. Similarly, negative and statistically significant values outnumber positive and statistically significant values by a factor of almost 60 to 1. This clearly points toward the inferiority of the MA strategy relative to the BH strategy for large-cap US stocks. Next, the evidence for market-timing ability points toward the overall inability of the MA strategy to be on the right side of the market. Negative values for β_{m^2} outnumber positive values by a factor of 1.5 to 1 with roughly the same ratio when only statistically significant coefficients are considered. Similarly, negative values for γ_m outnumber positive values by a factor of 2.5 to 1 increasing to a factor of

4.5 to 1 when we focus only on the values that are statistically significant. Finally, negative abnormal returns outnumber positive abnormal returns by a factor of 1.5 to 1 increasing to a factor of 50 to 1 when we only consider statistically significant αs.

Next, Panel B of Table 5.1 presents a summary of the distribution of the same parameters of interest for mid-cap US stocks. It is worthwhile to compare the mean and median values of $\Delta\mu$ and BETC of the mid-cap stocks to those for the large-cap stocks. This confirms formally the cursory observation from Section 5.2 that the MA strategy performs even more poorly for mid-cap stocks than it does for large-cap stocks. The market-timing ability of the MA strategy for mid-cap stocks is also relatively poor as evidenced by the number of negative market-timing coefficients, especially those that are statistically significant. Furthermore, the distribution of abnormal returns is even more negatively skewed for mid-cap stocks than it is for large-cap stocks. The number of negative abnormal returns exceeds the number of positive abnormal returns by a factor of 8 to 1 increasing to a factor of 15 to 1 when we only look at the statistically significant αs.

Finally, Panel C of Table 5.1 lists the distribution of the performance parameters of the MA strategy for small-cap US stocks. Comparing these values across Panels A and B it becomes clear that the MA strategy has the worst performance for small-cap stocks than it does for either large-cap or mid-cap stocks. The mean and median values for $\Delta\mu$ and BETC are in this case even more negative indicating that the MA strategy underpeforms the BH strategy by more with small-cap stocks than it underperforms with either large-cap or mid-cap stocks. Similarly, the number of negative abnormal returns outnumbers the number of positive abnormal returns by a larger factor than in the case of large-cap and mid-cap stocks. The only mitigating evidence here is with respect to the market-timing coefficients. The overall evidence does point toward poor market-timing ability but less strongly than for large-cap and mid-cap stocks as evidenced by the relatively larger number of positive values of the two market-timing coefficients compared to the number of negative values. This appears to be case when we consider only statistically significant values.

The empirical findings in this chapter are largely at odds with the empirical findings presented in previous chapters where we considered portfolios of US stocks rather than individual US stocks. This is quite puzzling given the fact that the US stock portfolios are constructed and value-weighted with the same individual US stocks we investigated in this section. This is one puzzle that merits further investigation and is left for future work.

Table 5.1 Summary statistics for US Stocks

This table reports the mean, median, percent positive, and percent negative as well as statistically significantly positive and negative $\Delta\mu$, BETC, β_{m^2}, γ_m as well as α from Carhart four-factor regressions for three sets of US stocks. The alphas are annualized and in percent. The sample period covers the first available day with a non-missing stock return for each stock until December 31, 2013. The length of the MA window is 20 days. Newey and West (1987) standard errors with 5 lags are used in reporting statistical significance. %PS indicates percent positive and statistically significant, %P indicates percent positive, %N indicates percent negative, and %NS indicates percent negative and statistically significant.

	$\Delta\mu$	BETC	β_{m^2}	γ_m	α
	Panel A: Large-capitalization US stocks				
Mean	−11.57	−0.0160	−0.0049	−0.0658	−0.0236
Median	−10.07	−0.0136	−0.0048	−0.0497	−0.0181
%PS	1.24	1.24	19.88	8.07	0.41
%P	3.73	3.73	38.51	30.43	48.45
%N	96.07	96.07	61.28	69.36	76.19
%NS	67.78	67.78	39.54	36.44	25.26
	Panel B: Mid-capitalization US stocks				
Mean	−14.85	−0.0242	−0.0045	−0.0545	−0.0355
Median	−12.52	−0.0196	−0.0058	−0.0578	−0.0258
%PS	2.89	2.89	17.11	9.63	2.14
%P	6.68	6.68	36.10	33.96	10.70
%N	93.32	93.32	63.90	66.04	78.34
%NS	49.13	49.13	37.70	36.36	31.02
	Panel C: Small-capitalization US stocks				
Mean	−22.87	−0.0395	−0.0020	−0.0440	−0.0636
Median	−18.62	−0.0321	−0.0028	−0.0429	−0.0475
%PS	1.37	1.37	18.39	10.51	0.35
%P	4.55	4.55	43.96	38.70	1.58
%N	95.45	95.45	56.04	61.30	84.41
%NS	29.76	29.76	26.80	27.32	40.28

Chapter 6

Concluding Remarks

There is overwhelming evidence that the switching MA strategy dominates in a mean–variance sense buying and holding any of the decile portfolios. The excess returns of the switching MA returns over buying and holding the underlying portfolios are relatively insensitive to the four Carhart (1997) factors and generate high statistically and economically significant alphas. In addition, abnormal returns for most deciles decline substantially after controlling for the dividend yield on the market portfolio, recessions, and up/down markets. This switching strategy does not involve any heavy trading when implemented with daily returns and very often has positive BETC, suggesting that it will be actionable for institutional investors. The findings are robust with respect to portfolio construction, various lag lengths of the MA, and alternative sets of portfolios. Last but not least, the lagged signal indicating whether the price has crossed the simple MA has substantial predictive power over the subsequent index return controlling for standard predictive variables mentioned previously. The risk-adjusted performance disappears only in the context of market-timing regressions in the framework of Henriksson–Merton (1981) where the downside market return is included as an additional factor and empirical asset pricing models with macroeconomic state variables. Hence, it appears that the success of the MA strategy does not represent an anomaly and is consistent with rational asset pricing. In addition, any abnormal returns surviving the previously mentioned tests may not be actionable in practice due to limits to arbitrage and price impact of trading on illiquid risky assets with low trading volumes.[1] The evidence for individual US stocks is quite the opposite to the one for US portfolios. It appears that using an MA strategy with individual US stocks leaves an investor worse off than the buy-and-hold strategy. This evidence holds for large-cap stocks, mid-cap stocks, as well as small-cap stocks. The indicators of successful market timing in both regression frameworks indicate that the MA strategy fails to time the market for individual US stocks. Finally, most of the four-factor alphas are negative and statistically significant indicating that the MA strategy with individual US stocks underperforms relative to the standard Carhart (1997) asset pricing model.

A more aggressive implementation involving selling short the underlying asset in response to a signal to switch instead of shifting the funds into cash results in higher returns but yields a much smaller risk reduction. The payoff of this version of the MA strategy resembles an imperfect at-the-money straddle. A further alternative involving investing in cash rather than in the risk-free asset produces inferior average returns, as expected. Finally, skipping a full day before trading does reduce the average return of the MA strategy relative to the buy-and-hold strategy.

It would also be of use to test more formally whether higher moments like skewness and kurtosis are improved by the MA strategy over the BH strategy. One potential alternative is to combine all first four moments using a utility function over them and convert the gains into certainty equivalent utility gains. Comparing those gains to the BETC will provide further evidence into the superiority of the MA switching strategy.

Considering the vast literature on technical analysis and the numerous technical indicators followed by some traders in practice, this study is just a first step toward investigating the performance and implementation of one common technical indicator. Future work will determine which other technical indicators perform well and whether they produce significant abnormal returns over and above the relevant transaction costs.

Notes

1 Fundamental versus Technical Analysis

1. Further findings using cash-flow-to-price, earnings-to-price, dividend-price, past return, and industry are broadly consistent with those reported in the text and are available from the author upon request.

2 Investment Performance

1. I use value-weighted portfolio returns to control for the amount of rebalancing trading inside the various portfolios. The empirical results in this paper are much stronger when equal-weighted portfolios are used. However, this may understate the BETC as equal-weighted portfolios require a lot of trading to be replicated.
2. I am very grateful to an anonymous referee for rasing the question regarding my motivation for using this particular length for the MA strategy.
3. Issues related to the statistical significance of the mean return improvement and the return standard deviation reduction are explored in the next section.
4. A quick glance at Table 2.1 reveals that the Low momentum (extreme loser) portfolio has the highest σ as well as the highest $\Delta\mu$ suggesting that it might be an outlier. Dropping this observation from the cross-sectional regression reduces the magnitude of the slope coefficient to 0.16 but it is still statistically and economically significant.

4 Performance Sensitivity

1. The robustness checks presented here are only a small portion of the total number of robustness checks performed in preparing this article. Results for equal-weighted portfolio, both daily and monthly returns, double-sorted portfolio sets along size/book-to-market and size/past performance show the profitability of the MA switching strategy is robust with respect to

the frequency of the data, the portfolio construction, and the portfolio composition. Additional results are available from the author upon request.
2. The full regression results along with the factor loadings are available from the author upon request.

5 Individual Securities

1. The remaining 17 components had price histories that were deemed too short to be included in the analysis of the comparative investment performance of the MA strategy with the BH strategy.
2. Having *t*-statistics that are, roughly, less than −2.
3. Having *t*-statistics that exceed 2.
4. The remaining 26 components had price histories that were deemed too short to be included in the analysis of the comparative investment performance of the MA strategy with the BH strategy.
5. The remaining 29 components had price histories that were deemed too short to be included in the analysis of the comparative investment performance of the MA strategy with the BH strategy.

6 Concluding Remarks

1. A variant of the moving strategy using stock futures and interest rate futures (instead of trading the stock and the risk-free asset) could address this point in practice. I leave the study of this version of the MA for future investigation.

Bibliography

Allen, F. and Karjalainen, R., 1999, "Using Genetic Algorithms to Find Technical Trading Rules," *Journal of Financial Economics* 51, 245–272.
Baker, M. and Wurgler, J., 2006, "Inevestor Sentiment and the Cross-Section of Stock Returns," *Journal of Finance* 61, 259–299.
Baker, M. and Wurgler, J., 2007, "Investor Sentiment in the Stock Market," *Journal of Economic Perspectives* 21(2), 129–151.
Balduzzi, P. and Lynch A. W., 1999, "Transaction Costs and Predictability: Some Utility Cost Calculations," *Journal of Financial Economics* 52, 47–78.
Barberis, N., Schleifer, A. and Vishny, R., 1998, "A Model of Investor Sentiment," *Journal of Financial Economics* 49, 307–343.
Bessembinder, H. and Chan, K., 1998, "Market Efficiency and the Returns to Technical Analysis," *Financial Management* 27(2), 5–17.
Black, F. and Scholes, M., 1973, "The Pricing of Options and Corporate Liabilities," *Journal of Political Economy* 81, 637–654.
Blume, L., Easley, D. and O'Hara, M., 1994, "Market Statistics and Technical Analysis: The Role of Volume," *Journal of Finance* 49, 153–181.
Brock, W., Lakonishok, J. and LeBaron, B., 1992, "Simple Technical Trading Rules and the Stochastic Properties of Stock Returns," *Journal of Finance* 47, 1731–1764.
Brown, D. P. and Jennings, R. H., 1989, "On Technical Analysis," *The Review of Financial Studies* 2, 527–551.
Campbell, J. Y. (1987), "Stock Returns and the Term Structure," *Journal of Financial Economics* 18, 373–399.
Carhart, M. M., 1997, "On Persistence in Mutual Fund Performance," *Journal of Finance* 52(1), 57–82.
Cochrane, J. H., 2008, "The Dog That Did Not Bark: A Defense of Return Predictability," *Review of Financial Studies* 21, 1533–1575.
Daniel, K., Hirshleifer, D. and Subrahmanyam, A., 1998, "Investor Psychology and Security Market Under- and Overreactions," *Journal of Finance* 53, 1839–1885.
Faber, M. T., 2007, "A Quantitative Approach to Tactical Asset Allocation," *Journal of Wealth Management* 9(4), 69–79.
Fama, E. F. and Blume, M. E., 1966, "Filter Rules and Stock-Market Trading," *Journal of Business* 39, 226–241.
Fama, E. F. and French, K. R., 1992, "The Cross-Section of Expected Stock Returns," *Journal of Finance* 47(2), 427–465.
Fama, E. F. and Schwert, W., 1977, "Asset Returns and Inflation," *Journal of Financial Economics* 5, 115–146.
Ferson, W. E. and Schadt, R. W., 1996, "Measuring Fund Strategy and Performance in Changing Economic Conditions," *Journal of Finance* 51, 425–461.

Gencay, R., 1998, "The Predictability of Security Returns with Simple Technical Trading Rules," *Journal of Empirical Finance* 5, 347–359.

Goh, J., Jiang, F., Tu, J. and Zhou, G., 2012, "Forecasting Government Bond Risk Premia Using Technical Indicators," Working paper.

Han, Y., 2006, "Asset Allocation with a High Dimensional Latent Factor Stochastic Volatility Model," *Review of Financial Studies* 19, 237–271.

Han, Y., Yang, K. and Zhou, G., 2013, "A New Anomaly: The Cross-Sectional Profitability of Technical Analysis," *Journal of Financial and Quantitative Analysis* 48(5), 1433–1461.

Han, Y. and Zhou, G., 2013, "Trend Factor: A New Determinant of Cross-Section Stock Returns," Working Paper.

Henriksson, R. D. and Merton, R. C., 1981, "On Market Timing and Investment Performance. II. Statistical Procedures for Evaluating Forecasting Skills," *Journal of Business* 54(4), 513–533.

Hong, H. and Stein, J. C., 1999, "A Unified Theory of Underreaction, Momentum Trading, and Overreation in Asset Markets," *Journal of Finance* 54, 2143–2184.

Hsu, P.-H. and Kuan, C.-M., 2005, "Reexamining the Profitability of Technical Analysis with Data Snooping Checks," *Journal of Financial Econometrics* 3, 606–628.

Huang, D. and Zhou, G., 2013, "Economic and Market Conditions: Two State Variables that Predict the Stock Market," Working Paper.

Jensen, M. C. and Benington, G. A., 1970, "Random Walks and Technical Theories: Some Additional Evidence," *Journal of Finance* 25, 469–482.

Jiang, F., 2013, "Trend-Based Conditional Asset Pricing: Explaining the Cross-Section of Technical Analysis Profitability," Working Paper.

Kilgallen, T., 2012, "Testing the Simple Moving Average across Commodities, Global Stock Indices, and Currencies," *Journal of Wealth Management* 15(1), 82–100.

Lo, A., Mamaysky, H. and Wang, J., 2000, "Foundations of Technical Analysis: Computational Algorithms, Statistical Inference, and Empirical Implementation," *Journal of Finance* 55, 1705–1765.

Lynch, A. W. and Balduzzi, P., 2000, "Predictability and Transaction Costs: The Impact on Rebalancing Rules and Behavior," *Journal of Finance* 66, 2285–2309.

Malkiel, B. G., 1996, A Random Walk Down Wall Street, W. W. Norton, New York.

Merton, R. C., 1981, "On Market Timing and Investment Performance. I. An Equilibrium Theory of Value for Market Forecasts," *Journal of Business* 54, 363–406.

Neely, C. J., Rapach, D. E., Tu, J. and Zhou, G., 2010, "Out-of-Sample Equity Premium Prediction: Fundamental vs. Technical Analysis," Unpublished working paper, Washington University in St. Louis.

Neely, C. J., Rapach, D. E., Tu., J. and Zhou, G., 2011, "Forecasting the Equity Risk Premium: The Role of Technical Indicators," Unpublished working paper, Federal Reserve Bank of St. Louis.

Neely, C., Weller, P. and Dittmar, R., 1997, "Is Technical Analysis in the Foreign Exchange Market Profitable? A Genetic Programming Approach," *Journal of Financial and Quantitative Analysis* 32, 405–426.

Neftci, S. N., 1991, "Naive Trading Rules in Financial Markets and Wiener-Kolmogorov Prediction Theory: A Study of 'Technical Analysis'," *Journal of Business* 64, 549–571.

Newey, W. K. and West, K. D., 1987, "A Simple, Positive Semi-Definite, Heteroscedasticity and Autocorrelation Consistent Covariance Matrix," *Econometrica* 55, 703–708.

Pastor, L. and Stambaugh, R., 2003, "Liquidity Risk and Expected Stock Returns," *Journal of Political Economy* 111, 642–685.

Ready, M. J., 2002, "Profits from Technical Trading Rules," *Financial Management* 31, 43–61.

Sharpe, W. F., 1964, "Capital Asset Prices: A Theory of Market Equilibrium under Conditions of Risk," *Journal of Finance* 19(3), 425–442.

Sullivan, R., Timmerman, A. and White, H., 1999, "Data-Snooping, Technical Trading Rule Performance, and the Bootstrap," *Journal of Finance* 54, 1647–1692.

Sweeney, R. J., 1988, "Some New Filter Rule Tests: Methods and Results," *Journal of Financial and Quantitative Analysis* 23, 285–300.

Taleb, N. N., 2012, Antifragile Things that Gain from Disorder, Random House, New York.

Treynor, J. L. and Ferguson, R., 1985, "In Defense of Technical Analysis," *Journal of Finance* 40, 757–773.

Treynor, J. L. and Mazuy, K., 1966, "Can Mutual Funds Outguess the Market?", *Harvard Business Review* 44, 131–136.

Zhang, F. X., 2006, "Information Uncertainty and Stock Returns," *Journal of Finance* 61, 105–136.

Zhu, Y. and Zhou, G., 2009, "Technical Analysis: An Asset Allocation Perspective on the Use of Moving Averages," *Journal of Financial Economics* 92, 519–544.

Index

9781137364685